"In his compelling history of the rover's place in the space program, *Across the Airless Wilds*, Earl Swift writes that, during Apollo 15, 16, and 17, astronauts drove it over 56 miles. . . . Such are Mr. Swift's narrative talents and the bounties of the source material that the book is a joy to read from beginning to end. . . . Swift has reminded readers of an endlessly fascinating chapter in space exploration with widespread implications for the future."

—*Wall Street Journal*

"Swift relays the awe-inspiring story of Apollo 17 and the lunar vehicle in a way that makes it all feel brand new. From the sheer human ingenuity of the moon missions to the deeply human figures inside the space suits, this book is a brilliantly observed homage to the human spirit."

—*Newsweek*

"The literature of lunar exploration has tended to focus on the earlier Apollo missions, with scant attention paid to the extraordinary achievements of the later rover expeditions—which were, in many respects, scientifically bolder and taught us a great deal more about our moon. Earl Swift lays out this great unsung saga with verve and magisterial sweep. After reading *Across the Airless Wilds*, you'll begin to think of NASA's true golden age not in terms of 'one small step,' but as a series of cosmic car rides."

—Hampton Sides, author of *In the Kingdom of Ice*

"Such an enjoyable book. . . . A clear and compelling story. . . . Ingenious. . . . Up-end[s] the Apollo narrative entirely so that every earlier American venture into space was preparation for the last three trips to the moon. . . . [A] detailed, thrilling account. . . . Swift conveys the baffling, unreadable lunar landscape very well, but *Across the Airless Wilds* is above all a human story, and a triumphant one at that."

—*Times* (London)

"With *Across the Airless Wilds* Swift has illuminated an underappreciated marvel of engineering and created a fabulous summer read. The entire second half of his book is a riveting travelogue of six astronauts' 'lunar road trips' drawn from interviews and radio transcripts, augmented by priceless color photos that alone are worth the price of the book."
—*Virginian-Pilot*

"In *Across the Airless Wilds,* Earl Swift skillfully tells the remarkable story of how vision, ingenuity, and some pretty fine engineering transformed lunar and planetary exploration. A rare and compelling celebration of the human spirit."
—Andrew H. Knoll, professor of earth and planetary science at Harvard University; member of the science team of NASA's Mars Exploration Rover mission; and author of *A Brief History of Earth*

"This is not just a book about the lunar rover—it's also a book about humans, and the great things they can do when inspired. There are people here who jump off the page—and sometimes, off the moon's surface. Vividly written, engaging, and fascinating. I started it one day and finished it the next, and I'm not a fast reader. I just didn't want to stop."
—James Donovan, author of *Shoot for the Moon: The Space Race and the Extraordinary Voyage of Apollo 11*

"For the origins and history of the Apollo lunar rover, there is no better guide than Earl Swift's beautifully written book. It details two decades of rover concepts, followed by two frantic years of building one for Apollo on a ridiculous schedule and an inadequate budget. But it paid off in three spectacular landings that used the rover for science—Apollos 15, 16, and 17. Swift also profiles the people who accomplished this feat; they are as fascinating as the machine itself."
—Michael J. Neufeld, Senior Curator, Space History Department, Smithsonian National Air and Space Museum, and author of *Von Braun: Dreamer of Space, Engineer of War*

ACROSS THE
AIRLESS WILDS

Chesapeake Requiem: A Year with the Watermen of Vanishing Tangier Island

Auto Biography: A Classic Car, an Outlaw Motorhead, and 57 Years of the American Dream

The Big Roads: The Untold Story of the Engineers, Visionaries and Trailblazers Who Created the American Superhighways

Where They Lay: Searching for America's Lost Soldiers

Journey on the James

The Tangierman's Lament and Other Tales of Virginia

One Buck at a Time: An Insider's Account of How Dollar Tree Remade American Retail (with Macon Brock)

ACROSS THE AIRLESS WILDS

The Lunar Rover and the Triumph
of the Final Moon Landings

EARL SWIFT

MARINER BOOKS

Boston   New York

HarperCollins books may be purchased for educational, business, or sales promotional use. For information, please email the Special Markets Department at SPsales@harpercollins.com.

A hardcover edition of this book was published in 2021 by Custom House, an imprint of William Morrow.

FIRST MARINER PAPERBACK EDITION PUBLISHED 2022.

Designed by Paula Russell Szafranski
Half title and title pages images © NASA

Library of Congress Cataloging-in-Publication Data

Names: Swift, Earl, 1958– author.
Title: Across the airless wilds : the Lunar Rover and the triumph of the
    final moon landings / Earl Swift.
Description: First edition. | New York, NY : Custom House, [2021] |
    Includes bibliographical references and index.
Identifiers: LCCN 2021006863 (print) | LCCN 2021006864 (ebook) | ISBN
    9780062986535 (hardcover) | ISBN 9780062986542 (trade paperback) |
    ISBN 9780062986559 (ebook)
Subjects: LCSH: Lunar surface vehicles—United States—History—20th
    century. | Project Apollo (U.S.) | Space flight to the moon.
Classification: LCC TL480 .S95 2021 (print) | LCC TL480 (ebook) | DDC
    629.45/4—dc23
LC record available at https://lccn.loc.gov/2021006863
LC ebook record available at https://lccn.loc.gov/2021006864

ISBN 978-0-06-298654-2

22 23 24 25 26 LSC 10 9 8 7 6 5 4 3 2 1

For Gerry Swift

CONTENTS

PART ONE

THE DIFFERENCE IT MADE

THE U.S. SPACE AND ROCKET CENTER ANNOUNCES ITSELF FROM miles away, with a needle against the sky that orients, at a glance, anyone in Huntsville, Alabama: if you can see the Saturn V, you can place where you are.

The rocket towers 363 feet over an especially smart precinct of a smart city in a state largely uncelebrated for its smarts. Just to the north lies a University of Alabama campus big on science and engineering. Clustered nearby are dozens of high-tech companies doing business behind locked doors and security cameras. South of the Saturn V lies the magnet for this brainpower: the George C. Marshall Space Flight Center. The place that produced the rockets that carried America to the moon.

The Saturn is a well-executed fake, erected in 1999 to mark the thirtieth anniversary of the first lunar landing. The U.S. Space and Rocket Center is not part of the National Aeronautics and Space Administration, the federal agency that achieved that landing; it's a museum operated by the state of Alabama. But it's a good one, with an impressive collection of genuine space hardware and a world-famous Space Camp for aspiring astronauts, and with one of the earth's three surviving real Saturn Vs on display inside.

I pulled into town one Wednesday in April 2019, spied the needle on the distant horizon, and followed it to the Space and Rocket Center. The mock Saturn had just received fresh paint in preparation for the fiftieth anniversary of that first moon landing, and the museum's gift shop was stocked full of T-shirts, ball caps, coffee mugs, and toys commemorating July 20, 1969. It being a school day, only a few customers browsed the shop. Most were older than me—not surprising, perhaps, as I could recall little about Apollo 11; I remembered my parents' excitement over the landing more than the event itself. It wasn't until the last few Apollo missions that I paid much mind to what was happening up there. By then, I was a teenager, and some mornings read a newspaper. Plus, we'd moved to Houston, home to the astronauts and NASA's center for manned spaceflight, and my eighth-grade classmates actually discussed lunar exploration.

But I remembered those later missions, too, for a distinction that set them apart, a new piece of gear the astronauts of Apollos 15, 16, and 17 carried with them. An addition that had redefined lunar exploration, space science, and NASA's expectations of what could be achieved in brief visits to the moon's inhospitable surface. The gift shop had no wares memorializing that transformative hardware, but I happened to know I could find it on display in the museum proper. Which is why I had driven eight hours to Huntsville: to see it in person, and to meet a man central to its creation.

My ticket bought me into the museum's centerpiece building, the Davidson Center for Space Exploration. Its main floor is a single cavernous room, 476 feet long, 90 wide, and six stories high, down the middle of which runs its main exhibit. The Saturn V is displayed on its side and broken into its component stages to show off the engines on each. Three old-timers were seated together on a bench under the rocket's enormous F1 engines, which jutted from the bottom of its S-IC booster stage. They are the strongest rocket engines ever put to use, and amid the museum's bright, kid-friendly cheer, the brutal, elemental power manifest in their fat tangles of pipes, valves, and pumps was unnerving. Their fluted mouths, a dozen feet across and built to spout great tails of fire and thunder, were no less fearsome for their silence. The men on the bench were inured to the menace overhead. All wore white lab coats that identified them as museum docents and retired rocket scientists.

I stopped in front of the bench. They were in their eighties, by the looks of them, maybe older. "Excuse me," I said. "I'm looking for Sonny Morea. Do you know where I could find him?"

"He was just here," one of the men replied. On the right breast of his lab coat, beneath his name tag, was an embroidered logo bearing the legend "NASA Emeritus."

"He's here," another said. "He's around. He might have just stepped away for a minute."

"He might be down there," the first man told me, pointing to the Davidson Center's far end. "That's where he usually is, back in that corner."

I thanked them and started that way, walking beneath the Saturn, which rests well off the floor on heavy steel cradles. Its scale borders on the absurd. The S-IC stage, essentially a flying gas can that muscled all 6.2 million pounds of the rocket off the pad and into the upper atmosphere, is 138 feet long and 33 feet in diameter; it occupies half the Davidson's lofty headspace lying down. The S-II stage, which took the Saturn into the airless black, came after, just as big around and 82 feet long. I passed under the S-IVB, or third stage—narrower, at 21 feet, 8 inches across, but still a monster. It put the astronauts into orbit, sent them on their way to the moon, and carried, in a shroud at its top, the lunar module. Beyond was a mock-up of the comparatively tiny Apollo spacecraft, the payload for all the rocket below. The main act, it consisted of the service module (supplying power, air, water, and electronics to the crew) and the command module, otherwise known as the capsule.

It took me several minutes to walk the length of this behemoth, with requisite pauses to admire its audacity. Stacked and ready for lift-off, it had stretched more than three times the length of the Wright brothers' first flight. Up under the capsule's nose, four hundred feet from the old-timers at the tail, I saw that they'd been joined by a fourth figure in a lab coat. I hurried back to introduce myself.

I'd seen photographs of Saverio "Sonny" Morea taken in the late 1960s—nattily turned out, by the professional engineering standards of the day, in crisp oxford shirts, bow ties, and skinny-lapeled sport coats, his dark hair trimmed short on the sides in the prevailing

NASA style. The photos hinted at a certain consistency of temperament: Whenever a flash went off, Sonny Morea seemed to wear an expression of expectant delight. They suggested that here was a guy who enjoyed conversation, liked people, didn't sweat the little stuff.

Fifty years had passed since—he was now eighty-seven—but it was instantly clear that Morea's long exposure to Earth's gravity had done little to mask his cheer. We shook hands. "I'm sorry I wasn't here when you got here," he said, smiling. "Have you been down there to see it?" He had a crooner's tenor still tinged with Richmond Hill, Queens, sixty-odd years after he left the old neighborhood—and with almost all of that time spent in Alabama, no less. "No," I told him. "I didn't get that far."

"Well, then, let's go have a look."

Back up the length of the Saturn we ventured, Morea tilted forward about ten degrees and hurrying on very short steps to keep up with the lean. Past the rocket's tip, the Davidson's gallery was twilit, with spots illuminating a few Holy Grail items of Apollo history: the scorched command module, *Casper*, from Apollo 16, interior lit to show off the three "couches" its crew occupied; an A7L space suit, worn in Earth orbit and now encased in glass; and what we'd come to see.

The lunar rover—in NASA parlance, the lunar roving vehicle, or LRV—was just beyond reach behind a low barrier. The "moon buggy," as the press insisted on calling it when it carried the astronauts around in the early seventies. A "spacecraft on wheels," as Morea and his fellow engineers preferred to think of it. "There it is," he said now. "What do you think?"

What I thought was that it looked just as I'd imagined it. All business. Built with the precision and purpose of all Apollo machinery. A wondrous meld of engineering and imagination, deceptively simple, conceived at a time when the available tools to work out its hidden complexities were slide rules, blackboards, and hand-drawn blueprints. "Wow," I answered. "Amazing." And I meant it, because I understood that I was beholding something truly revolutionary. And elegant. And rare.

Even as I spoke, though, it occurred to me that the uninformed observer might be less impressed. Suspicious, even: It wasn't entirely clear that this rover was the real thing, because it was displayed with the Davidson Center's one unalloyed disappointment, a mock-up of the lunar module that looked salvaged from a high school stage production. That aside, there was the vehicle itself, insofar as there was very little to it—it was tiny, and its seats looked like lawn chairs, and it lacked a body or a roof or a steering wheel or much of anything, besides wheels, that typically define a car. Next to the Saturn V reclining big as creation a few feet away, it looked like a weekend garage project abandoned well before its finish.

But it takes imagination to celebrate what's missing in an object, along with what's there, and the LRV was a feat of "less is more" engineering that was radical even by NASA's standards. Its builders distilled everything essential to an earthbound off-roader to its indivisible minimum, its smallest and lightest and most fundamental iteration, then whittled even further. On Earth, it weighed 460-odd pounds—not much more than a single astronaut in his space suit—but on the lunar surface weighed a sixth as much, thanks to the moon's weaker gravity. So it was that the four electric motors turning its wheels together churned out just one horsepower. You can buy a gutsier shop vac, but at less than eighty lunar pounds, that's all the punch it needed.

On the job, it proved sturdy enough to shrug off a lot of abuse. Yet the museum's example rested on a display stand that propped up its aluminum frame; without it, the chassis might have sagged under the pull of Earth's gravity by now. If I were to step onto the floorboard, I'd snap it. I focused on its tires, which I'd long admired in pictures. They were the shape and size of all-season radials, only made of stainless steel mesh. They supported the rover on the moon, but here, minus the stand, would squash flat. Each, rim included, weighed just twelve pounds. Earth pounds.

"It doesn't look like much," Morea allowed. "It weighs next to nothing. But, you know, I worked on the Saturn for ten years before we started this. I worked on the F1 engines—I managed that program for seven years—and then I was sent in to help fix the J2 engines." He turned to the rocket and nodded toward the fonts of hellfire on the second and third stages. "Couldn't go anywhere without those. But it's kind of crazy: of all the things I worked on, this is what I'll probably be remembered for."

I eyed the rover. "People can't get their heads around rocket engines," I suggested. "I look at one, and I can't figure it out. This, I understand."

"Yeah," Morea said, nodding. "Who hasn't driven a car?"

The machine seemed familiar enough fifty years ago that some of the press treated it as the inevitable, almost comic product of the most automotive people on Earth. Of *course* we'd send a car into space.

In truth, there was no "of course" about it. Basic layout aside, the rover had little in common with any other vehicle built in the nearly eighty years of the horseless carriage that preceded it, and bore no resemblance to any other 1969 General Motors product, which is essentially what it was. It was called on to cross country that no Earth car would encounter, in conditions that would cripple any terrestrial vehicle instantly: temperatures of minus 250 degrees Fahrenheit in the shade and plus 250 in the sunshine; a surface of clingy, abrasive dust that could foul any moving part; fierce solar radiation; and a constant shower of micrometeoroids smaller than grains of sand but moving faster than bullets. All while wrapped in an airless vacuum.

Under the circumstances, mere survival would have set the rover apart. But it also carried twice its weight, despite its gossamer construction. And endured lengthy odysseys on which it climbed slopes that would test any jeep, clawed over foot-high rocks, thumped into and out of craters. And kept track of where it was on the lunar surface, so that if anything went wrong, it could point the fastest way back to the lunar lander. Not that anything could go wrong: operating nearly a quarter million miles from the nearest service station required the rover to be reliable, above all else, with redundancies for its major components. If one of its motors failed, it could run on the other three. If two motors failed, it could run on the remaining two. If three failed, well, all bets were off, but it might limp for a while on one. That it met the demands of lunar travel is remarkable even today, when our cars have become rolling computers, and our poor habits are sniffed out by sensors and alarms and automatic braking. But the rover was designed in 1969, when both automotive engineering and space technology relied on a far more primitive set of tools.

It evolved from ideas that are older still. Within NASA, the rover project is remembered not only for its success but for how quickly it gelled. The first machine was delivered to the Kennedy Space Center just seventeen months after the space agency awarded the contract to produce it, a fraction of the time Apollo hardware usually took. It had a long family history, however: The rover was shaped by nearly a decade of start-and-stop NASA studies into how best to explore the moon— inquiries that signaled just how big the agency was thinking at the time and also assumed that Apollo would be the first chapter in a long lunar campaign. The vehicles conceived in those inquiries were left on paper, unbuilt, but traces survived in the little machine that Morea and I now admired. Alloyed into its metal, and especially those exotic wheels, were years of sweat, experimentation, and creativity.

When NASA finally went ahead with the project, hundreds of people raced the clock to get it aboard the Saturn V. Little went according to plan. The project blew through its budget and threatened to overshoot its deadline by months. Careers were made and broken in the struggle to finish it. The man standing beside me in the Davidson Center came close to calling the whole thing off, and NASA's higher-ups stood at the same precipice on numerous occasions.

Yet it got to the moon. In the face of myriad challenges it made it there and changed everything about Apollo.

2

CONSIDER THE TWO LUNAR MISSIONS OF 1971: APOLLO 14, WHICH landed in the moon's rugged Fra Mauro region in early February, and Apollo 15, which six months later took its astronauts to a plain rimmed by sky-high mountains and a mammoth canyon, and carried the first LRV.

On their second excursion from their lunar module *Antares*, Apollo 14 astronauts Alan B. Shepard Jr. and Edgar Mitchell embarked on the longest walk of the Apollo program—a hike over undulating ground to the rim of Cone Crater, a half mile away, where geologists hoped to find rocks from deep in the moon that had been blown from the hole during its creation. Pulling a two-wheeled, rickshaw-like contraption for their tools and rock samples, they set out confident that they knew just where the crater was—"right over that way," as Mitchell put it. Their maps depicted the landmarks they'd see on the way.

But the moon played tricks on them. The horizon was weirdly close. The sky was utterly black. The gray surface concealed its features behind swells and declivities. And the astronauts' perception of size and distance was jumbled by the absence of any visual yardsticks—trees or houses or clouds—so that a large rock hundreds of yards away looked no different from a smaller one close by. The same was true of craters: within minutes, the pair mistook small depressions for large, assigned them the wrong names, and in so doing, misjudged their location and thus their speed. When they first stopped to gather samples, they were hundreds of yards short of where they thought they were.

Not long after, they realized the surrounding moonscape didn't match their maps and disagreed on what they did see. Shepard pointed out what he took to be Weird Crater, a cluster of overlapping depressions off to their south. Mitchell thought they had to be "considerably past Weird." Mission Control in Houston, with only the astronauts' verbal descriptions to go on, had no way of knowing who was right, assuming that either was.

The dust underfoot, soft and yielding, began to tilt. "We're starting uphill now," Mitchell told Houston. "Climb's fairly gentle at this point, but it's definitely uphill." Moving in a pressurized space suit was taxing on flat ground; soon the slope had both astronauts breathing hard. An hour into the hike, they stopped to "take a break, get the map, and see if we can find out exactly where we are," as Mitchell said. As they caught their breath, they thought they sorted out their location. They were wrong.

Shepard and Mitchell took turns pulling the cart, which NASA— sweet on convoluting the labels for its gear and infatuated with acronyms—called a modular equipment transporter, or MET. Keeping it under control on the uneven ground slowed them. As the slope steepened, they picked up the MET and carried it, certain their climb would end at Cone Crater's rim. When they reached the top, however, they found a swale ahead and, beyond it, another rise. The crater, more than a thousand feet wide, was nowhere in sight. "Well," a baffled Shepard told Mission Control, "we haven't reached the rim yet."

"Oh boy," Mitchell said. "We got fooled on that one."

They marched on, pulling and carrying the MET, ascending another steep rise, growing more tired and frustrated by the minute, and steadily depleting the stores of air and cooling water in their backpacks. Their transmissions were breathless, and at times their heartbeats, especially the forty-seven-year-old Shepard's, spiked to the point that Houston urged them to rest. This climb, too, ended in disappointment. "It's going to take longer than we expected," Mitchell reported. "Our positions are all in doubt now."

They were not lost. They could see *Antares* behind them; getting back to safety was never in doubt. But experiments on Earth had shown that an impact crater's debris, or ejecta, is arranged around the hole in a predictable pattern, with the material from deepest underground closest in—and that if they wanted to sample lunar bedrock, which was a mission priority, they had to get near that rim. They came to another long slope, which Shepard figured would take thirty minutes to crest. "I don't think we'll have time to go up there," he told Mitchell.

"Oh, let's give it a whirl," his partner countered. "Gee whiz. We can't stop without looking into Cone Crater." They'd traveled a long way to get here, and they were close, wherever they were. He was confident they would "find what we're looking for up there."

Houston came on the radio. "In view of your assay of where your location is, and how long it's going to take to get to Cone, the word from the back room is they'd like you to consider *where you are* the edge of Cone Crater."

Mitchell, exasperated, called their handlers "finks," a mild putdown of the day. Shepard tried to assuage him. "I think what we're looking at right here, in this boulder field, Ed, is the stuff that's ejected from Cone."

"But not the lowermost part," Mitchell replied, "which is what we're interested in."

"Okay," Shepard said. "We'll press on a little farther, Houston."

Mission Control extended their allotted time for the expedition by thirty minutes. They trudged farther uphill. When the incline leveled out, they were not looking down into Cone, however, but just another shallow valley in Fra Mauro's wrinkled surface. By that time, they'd cut deep into their extra half hour. Houston had them sample the rocks and soil where they were, then start back.

Skip ahead to the last day of July, and Apollo 15's first traverse of the Hadley-Apennine region by lunar rover. Dave Scott and Jim Irwin had a much more ambitious assignment than their predecessors: to cross a mile of hummocky, cratered plain to a spectacular gorge called the Hadley Rille, then follow its edge to the foot of a mountain that, in sheer mass, rivaled the biggest massifs on Earth—and climb its side.

Thirteen minutes into their ride, they reached the rille, nearly a mile wide and a thousand feet deep. They'd already traveled twice the distance from *Antares* to the Cone Crater. And though piloting the rover involved some "sporty driving," as Scott told Houston, they remained fresh. They drove along the canyon's lip until they were two straight-line miles from their own lunar module, *Falcon,* and stopped alongside Elbow Crater, timeworn and eleven hundred feet wide, the site of their first geologic investigation.

Mission Control took remote control of the rover's TV camera. During Shepard's and Mitchell's slog at Fra Mauro, the camera back at *Antares* had stared out over motionless landscape; the only change in the picture was a slow, subtle shift in the scene's lighting as the sun crawled higher in the sky. Now, at Elbow Crater, viewers on Earth could watch Scott and Irwin in real time as they conducted science in the field, bagging samples and taking pictures.

They left for the mountain, Hadley Delta. The rover outperformed its meager horsepower: when the men stepped off the machine, 2.4 straight-line miles from base camp, they were surprised by how steep the ground felt under their feet and how high up the mountainside they'd climbed. They had a commanding view of the plain and canyon below, and for the moment, had driven beyond sight of their lunar module. "Oh, look at that," Scott said. "Isn't that something? We're up on a slope . . . and we're looking back down into the valley, and—"

"That's beautiful," Irwin said.

"That is spectacular," Scott agreed. He readied the rover's antenna for TV transmission, then paused for another long look. "The most beautiful thing I've ever seen."

They were rested and ready for work. Rides were comparatively effortless, with cooldowns built into every excursion. That yielded lower metabolic rates, which translated into slower consumption of the air and water in the astronauts' backpacks, and stretched the time they could spend outside. Scott and Irwin sampled rocks and soil for forty-five minutes before heading back downhill, following a course for base camp laid by the rover's navigation system. It worked

The Apollo 15 lunar roving vehicle, covered with dust but otherwise no worse for its more than seventeen-mile exploration of the moon. (NASA)

as designed, so that they always had a good fix on where they were. It also safeguarded them from one of the great disappointments at Fra Mauro: the Apollo 14 moonwalkers had come within sixty-five feet of Cone Crater's rim, but didn't know it.

3

THROUGHOUT HISTORY, WE HUMANS HAVE CELEBRATED THE FEW among us who have dared to go where others have not. Their names remain familiar long after their explorations: Erikson, Magellan, and Cook. Amundsen, Scott, and Shackleton. Peary and Henson. Livingstone. Columbus. Marco Polo. The twelve men who stepped onto the moon ventured farther from home, and faced a greater range of dangers, than all of them.

Yet we remember few of their names. Only the first pair to land are readily known to many (I dare not say most) Americans. That's understandable, if disappointing: Apollo 11's touchdown marked a triumph of imagination as much as technology. The courage it required, the precision it demanded, and the sheer boldness of the undertaking—not to mention the anticipation attending Neil Armstrong's first step onto the regolith—thrilled and inspired a world witnessing it live on television.

But fact is, the greatest achievements of our lunar adventure came later, when the world was no longer hanging on every word the moonwalkers spoke or following every step they took, on missions that are recalled dimly today. In fact, you could argue that every earlier American venture into space was preparation for the last three trips to the moon. The six manned Project Mercury flights of the early 1960s established that spacecraft and their passengers could survive the forces required to get them into Earth orbit and back. Ten manned Project Gemini missions sent up two astronauts at a time, demonstrated that they could function for days on end in space, and tested and refined the maneuvers and spacecraft dockings central to the moon missions to come.

Each of the early Apollo flights checked out the equipment and procedures necessary for a landing: In October 1968, the first manned mission, Apollo 7, test-drove the command and service module. Two months later, Apollo 8 fired humans into deep space and around the moon for the first time, aboard the first manned Saturn V; its crew became the first to witness the moon's far side and snapped Earth's most revealing selfie. Apollo 9 shook down the lunar module in low Earth orbit, while Apollo 10 served as a dress rehearsal for the coming first visit to the moon's surface.

Each of the early landings built on the former. Apollo 11 aimed simply to put its astronauts on the regolith and get them back alive: *Eagle* landed on the board-flat Sea of Tranquility, the safest but least interesting real estate on all the moon, and its crew didn't stray far: the footprints left by Armstrong and Buzz Aldrin would fit inside a football field, with a lot of yardage to spare. A few months later, Apollo 12 set down on another tame expanse of lunar desert, but within sight of an unmanned probe sent there two years before—which Charles

"Pete" Conrad and Alan Bean inspected in the course of hiking about 1.4 miles. They thereby established that a crew could land at a specific point, and cleared the way for more demanding, and interesting, destinations. Apollo 13 didn't have the chance to follow through on that promise, but Apollo 14 did, pinpoint-landing in the highlands of Fra Mauro. Shepard's and Mitchell's dispiriting march to the edge of Cone, long though it was, never took them much beyond a half mile from base.

Then, with Apollo 15, NASA applied all it had learned into putting its men and materiel to their best and highest use. Its exploits marked the program's transition to deeper science and far more swashbuckling exploration. In part, that was thanks to the beefed-up lunar module Scott and Irwin flew to the Hadley-Apennine, as well as the improved backpacks they wore, both of which granted them time for a longer look around. But more importantly, they had *range*.

"They were looking at how could the astronauts get the most bang for the buck—in getting around, in picking things up, in exploring," Morea told me in the Davidson Center, hands in the pockets of his lab coat, gazing at the rover with parental pride. "So, they looked at a number of ideas. One was a pogo stick." He glanced my way and chuckled. "Really, a pogo stick! But then, a car came up pretty fast."

A couple of seconds passed before he added: "Though it's not a car. It's really a spacecraft."

When they braked their rover to end that first excursion, two hours and sixteen minutes after its start, Scott and Irwin had covered 6.3 miles—more than all the travel achieved by the first three landing crews combined. All told, astronauts on the last three lunar visits drove more than *fifty-six miles*. Each mission's rovers could cover an area the size of Manhattan.

Ever bolder exploits followed. Sixteen months on, Apollo 17's LRV churned its way up a ridgelike fault that rose high above a lunar plain, then rolled down the other side. When astronauts Gene Cernan and Jack Schmitt stopped at the bottom to gather rock samples, they

Apollo 17 commander Gene Cernan stands at Nansen Crater in the Taurus-Littrow Valley, the farthest point of man's extravehicular wanderings on the moon. (NASA)

were nearly five miles from their lunar module, near the outer limit of their safe radius of travel. The moment the two climbed off their rover—at 8:36 P.M. EST on Tuesday, December 12, 1972—marked a pinnacle in the annals of exploration. No other explorer has been in circumstances so remote, or so extreme in their hazards. No expedition had before, or has since, pushed adventure farther or further. Cernan and Schmitt were out at the edge of the edge of man's travels as a species. By comparison, Roald Amundsen's trek to the South Pole was a run to the corner grocery.

As they drove, the rover crews piled their machine with moon rocks; of the 842 pounds of geologic samples collected on the six Apollo landings, nearly three-quarters, about 620 pounds, were gathered on drives in the LRVs. Shared with scientists around the globe, the samples have informed our understanding of the heavens ever since.

It comes to this: Remembered or not, the nine days the final three missions spent on the moon were a fitting culmination to Apollo, and a half century later remain the crowning accomplishment of America's manned space program. And their success would have been beyond reach without the wispy contrivance on display at the Davidson Center.

Or rovers just like it. The Davidson's machine was one of several built to test the design; this was the vibration test unit, constructed exactly like those certified for flight, but tormented for months in shakers, vacuums, ovens, and deep freezers to gauge its hardiness during launch and touchdown. It's as close to the genuine article as we can get: the three sent to the moon remain there, along with a scattered junkyard of other Apollo detritus.

Anyone who still doubts that astronauts visited the moon—and those people still walk among us, even at this late date—need only go online to find overhead photos of the landing sites, taken in 2011 by a lunar orbiter. In surprising detail, they show the three rovers parked near their landers.

Plainly visible all around them, and stretching for miles across the lunar wastes, are tire tracks.

PART TWO

NATION
OF
IMMIGRANTS

4

THOSE WITH A DIRECT ROLE IN THE ROVER'S CREATION NUMBERED in the hundreds: engineers and draftsmen, industrial designers, metallurgists and welders, geologists, soil scientists, and whiz kids in math, electronics, and the rudimentary computers of the day. Taken together, they represented a tiny but important sliver of the roughly four hundred thousand people who made contributions, great and small, to the Apollo program. Leading the hundreds were a few memorable characters. And this story starts, as stories about the space program tend to do, with one.

Wernher von Braun, familiar though his name might be, was an enigmatic figure throughout his long public life, and remains so more than forty years after his death. He was a man of bold vision, but selectively blind when it suited his ambition. He was a romantic dreamer, while also a pragmatist willing to sideline pesky ethics or empathy to achieve his goals. He was both a fastidious engineer and a handsome charmer who could work presidents and the press like an equation. And he was a card-carrying Nazi and SS officer—likely complicit in the deaths of thousands of concentration camp prisoners—whose signature achievement in World War II was a ballistic missile designed to kill Germany's foes. Whispered into the United States at war's end, he shape-shifted into a God-fearing, patriotic Alabaman who reimagined his missile to carry the Stars and Stripes to the moon.

Complicated doesn't begin to describe him. This can be said of von Braun without fear of contradiction, however: he imagined a future in space as few others had, sold the idea masterfully, and developed the hardware to make it real. As much as any other individual, he laid the groundwork for the U.S. space program, starting years before the

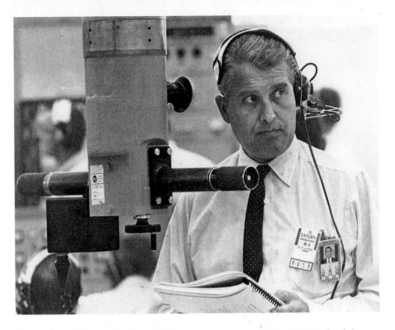

Wernher von Braun at a May 1964 Saturn launch, and as he approached the height of his career. (NASA)

first Mercury shots. And from early on, his dreams for the exploration of other worlds included vehicles zipping human adventurers and scientists across them.

He was born a Prussian aristocrat, Wernher Magnus Maximilian von Braun, in what is now Wyrzysk, Poland, in March 1912. His father was a cabinet-level officer of Germany's Weimar Republic, and von Braun was himself a Freiherr, or baron. He was a good-looking, winsome kid, with wavy blond hair, ice-blue eyes, and a gift for the arts. Early in his schooling, he mastered French and piano. He was uninspired by science and math.

But the year he turned thirteen, von Braun's parents marked his confirmation with a gift that launched a lifelong obsession. "I didn't get a watch and my first pair of long pants, like most Lutheran boys," he remembered as an adult. "I got a telescope." In time, he was spellbound by a magazine article about an imagined flight to the moon. "I don't remember the name of the magazine or the author," but the story "filled me with a romantic urge," he recalled. "Interplanetary travel! Here was a task worth dedicating one's life to! Not just stare through a telescope at the moon and the planets but to soar through the heavens and actually explore the mysterious universe! I knew how Columbus had felt."

This sense of destiny eventually led him to the writings of German-Romanian rocket pioneer Hermann Oberth, whose farsighted *The Rocket into Interplanetary Space* spelled out the science and engineering of space flight. The teenage von Braun found the slim volume a tough read—especially its reliance on complex math formulas—but was enthralled by its endorsement of liquid-fueled rockets (written before the first had flown) and its boldness. This was no magazine yarn. Oberth was a serious scientist, arguing for the real prospect of travel beyond Earth.

Inspired, von Braun dug deep into math and physics, bent on mastering both. And like other young Germans moved by Oberth's work, he sought membership in a rocket club, the Society for Space Travel. One of the club's founders, an avid rocketeer named Willy Ley, introduced the boy to Oberth, who was visiting from Romania. In the summer of 1930, under Oberth's tutelage, von Braun was part of a team that experimented with small rockets powered by alcohol and liquid oxygen. Oberth left for home, but not long after, in 1931, the group successfully launched one.

It wasn't the first such flight—American Robert Goddard had pulled off a liquid-fueled launch five years before—but the Germans didn't know that. Energized, the group moved its operations to an abandoned army munitions dump on the outskirts of Berlin, where it could safely tinker with more potent rockets, and von Braun evolved from apprentice mechanic, to field engineer, to leading thinker.

In retrospect, his rise seems almost preordained. He had it all—looks, a lightning-fast mind, and an infectious optimism and easy suavity that played well in any room. "Physically, he happened to be a perfect example of the type labeled 'Aryan Nordic' by the Nazis during the years to come," Ley said later. "His manners were as perfect as rigid upbringing could make them." He displayed a particular gift—a genius, even—for organization.

Von Braun was a full-time college student, and he and his fellow experimenters were next to broke. But their rockets, built from scrounged parts, their fuel tanks pressurized with bicycle pumps, flew well enough to attract the attention of the German army. In 1932, the year he left the Berlin Technological Institute for a doctoral program at the University of Berlin, von Braun accepted an invitation to continue his pursuit of rocketry at government expense. "We needed money, and the army seemed willing to help us," he explained later. "In 1932, the idea of war seemed to us an absurdity. The Nazis weren't yet in power. We felt no moral scruples about the possible future abuse of our brainchild. We were interested solely in exploring outer space." The following year, when Adolf Hitler took control, von Braun found himself the Third Reich's top civilian rocket man. He was twenty-one.

Von Braun's conduct over the succeeding dozen years has been the subject of so much scrutiny and controversy that here we'll tackle just the highlights. While earning his doctorate in physics at twenty-two, he laid the groundwork for the German rocket effort to come, organizing laboratories dedicated to the various facets of rocket development.

"I started out with one mechanic," he said later. "By the spring of 1937, the small development station at Kummersdorf Army Proving Grounds about thirty miles south of Berlin had grown to about eighty people. We successfully test-launched during this time two rudimentary liquid-fueled rockets . . . which reached altitudes of about a mile and a half."

That same spring of 1937, five years after starting work, he and his team moved the operation to Peenemunde, a resort town on the Baltic coast. The same year, he joined the Nazi Party—a political necessity, he'd explain later—and in 1940, the Schutzstaffel, or SS, in which he eventually held the rank of major. That didn't require explanation; it was scrubbed from his CV until years after he died. Meanwhile, Germany's leaders had precipitated World War II, and his efforts at Peenemunde focused on a superweapon: the world's first ballistic missile, a liquid-fueled, medium-range rocket that he called the A-4, but which is better known by the name that the Reich's propagandists hung on it—V-2, the *V* short for *Vergeltungswaffe*, or "vengeance." One October day in 1942, the team launched a prototype to an altitude of more than fifty miles, a stunning achievement. "We have invaded space with our rocket," declared General Walter Dornberger, von Braun's mentor and boss. "This third day of October, 1942, is the first of a new era, one of new transportation—that of space travel."

That assessment proved premature, because space travel didn't rank high in German priorities, especially as the country's fortunes crumbled in the war. Still, Peenemunde mushroomed in size and importance: a desperate Hitler came to see the V-2 as a potential game changer, and by late 1943, the operation employed thousands. Many were prisoners from the Dora-Mittelbau concentration camp, herded into an underground rocket factory to meet impossible production quotas. Conditions were horrific—thousands died of starvation, disease, and exhaustion, and scores of others were executed. Von Braun didn't resist employing slave labor and did nothing to mitigate the suffering that took place under his nose. His biographer, Michael J. Neufeld, surmises that he could have been prosecuted as a war criminal.

Luckily for him, the German secret police hauled him out of his house one night in March 1944 and tossed him into prison, ostensibly because von Braun and a couple of colleagues had been overheard bad-mouthing the war's trajectory and grousing that they'd much rather be firing rockets into space. More likely, it was payback for his resistance to an SS takeover of the V-2 project. Regardless, it took pleading from Dornberger and Albert Speer, the main architect of the German war effort, to win his release. Though the experience must have scared him, it also gave von Braun postwar credentials as a reluctant player in the Nazi war machine.

He would need them. On Friday, September 8, 1944, the first operational V-2s were fired from mobile launchers in Belgium and the Netherlands. One, sent on its way late that morning, landed in the southeast suburbs of Paris, killing six people and wounding thirty-six. The second was the opening shot in a seven-month campaign of terror. That missile rose in the early evening from Wassenaar, a genteel suburb of the Hague. It climbed fifty miles high in a steep, westward arc, then nosed over at the apogee of a parabolic course guided by gyroscopes and a rough analog computer. Minutes later, it reached its destination—Staveley Road in the west London neighborhood of Chiswick, a narrow residential lane bordered by cherry trees and lined with tidy duplexes of brick and stucco.

It came straight down at three times the speed of sound, so fast that it would have registered only as a white streak, a shooting star, to anyone who glanced up to see it. No one on Staveley Road had reason to look: the rocket outraced the sound of its approach, and arrived without so much as a sigh. In a flash, eleven houses were demolished, fifteen others staved and scorched beyond habitation, and scores damaged. A crater opened up in the concrete roadway, thirty feet wide by eight deep. For miles around, Londoners heard an odd double explosion that they'd come to know all too well, a sharp blast and a deep boom—the warhead exploding, followed by the gathered sound of a supersonic missile's last moments of flight.

In the rubble lay three bodies. Three-year-old Rosemary Clarke had been napping in the front bedroom of a flattened house. Ada Harrison, who ran a sweetshop with her husband, William, crawled from the wreckage next door to die in her yard. Army private Bernard Hammerton Browning walked into the eruption: he was on leave, and strolling to Chiswick Station to meet his girlfriend. At least seventeen people were injured, some critically. William Harrison was among them; he died in the hospital ten days later.

These were the first fruits of von Braun's labors for the Reich. All through the autumn and winter, the rockets kept coming. London and Antwerp, Belgium, were the German command's favored targets, but the V-2's guidance system was imprecise, so the surrounding countryside wasn't safe, either. Three golfers were seriously wounded in one of several strikes on British courses. People died over pints in rural pubs, and farmers in their fields. Casualties from most of the attacks were modest—two or three here, a half dozen there—but in the densely settled boroughs south and east of London's center, a few produced horrific tolls. The worst came on November 25, when a V-2 dropped into a Woolworth department store in New Cross at the height of its Saturday lunchtime trade, killing virtually everyone in the building, the occupants of a passing bus and truck, and shoppers lined up outside for the store's sale on tin saucepans. One hundred sixty-eight died, and dozens were gravely injured. One of the last of the missile strikes on Britain, on March 27, 1945, hit the Hughes Mansions, a cluster of brick, five-story apartment blocks in the city's Stepney district. It obliterated one, wrecked the others, and killed 134 men, women, and children.

As bad as things were in London, they were even worse in Antwerp. Just liberated, the city was an important port of entry for Allied troops and supplies, and more than 1,600 rockets were aimed at the place. One stands out. On December 16, 1944, more than a thousand people had crowded into the Rex Cinema to see the Saturday afternoon matinee of a Gary Cooper Western, *The Plainsman*, when a V-2 came through the roof. Of the 567 who died, 296 were Allied soldiers. Elsewhere in Antwerp, another 130 people were killed in rocket attacks that day. Hundreds more were wounded.

In all, more than three thousand V-2s were fired at Allied targets. They killed at least five thousand civilians and Allied combatants, and claimed many thousands more from the ranks who built them. Its creator was pleased with the V-2's performance, he'd recall after the war, except for one thing: it landed on the wrong planet.

By April 1945, the end was near, and the advancing Allies were on the prowl for the new technology. They poured into Germany from east and west seeking not only rocket hardware but the minds behind it: Tensions between the West and its communist partners were already sharpening, and both sides recognized the rocketeers as valuable assets in the postwar struggle for global power. At the same time, von Braun wagered that he and his team faced better futures if they surrendered to Americans instead of Soviets—"in the hope," as he'd later explain, "that we could come to the United States to pursue our real dream: spaceflight."

The group, including General Dornberger and a large contingent of Peenemunde scientists and engineers, fled to the highlands of southern Germany, where they holed up in a mountain lodge until U.S. troops occupied the countryside below. The GIs who took von Braun into custody could scarcely believe that here was the man behind the dreaded German rocket program. "There was considerable doubt in their minds as to whether he had ever launched anything more devastating than kites for the neighbors' Kinder," one recalled. "He seemed too young, too fat, too jovial." Dornberger helped change their minds. The general looked "splendidly evil," but "deferred humbly to von Braun, who, for his part, treated our soldiers with the affable condescension of a visiting congressman" and "conducted himself as a celebrity rather than a prisoner."

Days after the fighting in Europe ended, von Braun wrote a memo for his Allied captors, outlining the development of German rocketry and predicting the technology's future applications. A next logical step, he wrote, was multistage rockets that could achieve orbital velocities. From orbit, a rocket's crew could surveil the entire planet and even erect a permanent satellite base—a space station. "When the art of rockets is developed further, it will be possible to go to other planets," he wrote, "first of all to the moon."

5

WERNHER VON BRAUN'S SECOND ACT OPENED AT FORT BLISS, AN army post headquartered in El Paso, Texas, and spread across a vast swath of the Chihuahuan Desert. There, in the first year after the war, the Peenemunde team quietly set up house as part of Project Paperclip, a secret program to spirit German scientists and engineers into the States before the Soviets or anyone else got to them.

Their new home was board flat, sun blasted, and studded with cactus and saltbush. "Our first year here was a period of adjustment and professional frustration," von Braun would remember. "Distrusted aliens in a desolate region of a foreign land, for the first time we had no assigned project, no real task. Nobody seemed to be much interested in work that smelled of weapons, now that the war was over. And 'spaceflight' was a word bordering on the ridiculous."

The Germans were not idle, however. The army put some to work refining and test-firing confiscated V-2s at New Mexico's White Sands Proving Ground, which abutted Fort Bliss to the north. Some experimented with an early cruise missile. "In addition to rocketry," von Braun said, they "studied the American language, American government, and the American way of life." Von Braun himself was especially busy, becoming a born-again Christian, marrying his much younger first cousin, and moving her from Germany to Texas, along with his parents. He even found time to write a novel. *The Mars Project* failed for years to find a U.S. publisher, but it showcased its author's precise calculation of what would be required for a Mars voyage, starting with an immense, multistage rocket.

In time, he saw the concept take embryonic form. Throughout 1948, a White Sands team of army and General Electric engineers worked on Project Bumper—the first staged space vehicle, a V-2 modified to carry a much smaller WAC Corporal rocket on its nose.

The V-2 would serve as a booster, carrying the combined rocket from the launch pad into the high atmosphere, where the smaller rocket would fire and tear higher and faster until its tanks ran dry. In February 1949 a Bumper became the first object fired into outer space.

A few months after this triumph, the army decided to move its missile works from Fort Bliss to a shuttered chemical weapons arsenal and ordnance plant in Huntsville, Alabama. So, in 1950, von Braun and his Germans, along with several hundred civil service and military workers, picked up and moved east. Even the sequestered Germans knew something of Alabama's reputation for resisting change, to put it mildly, and they arrived with trepidation. "We knew that the people here run around without shoes," one of them, Konrad Dannenberg, recalled decades later. "They make their money moonshining, and that's what they drink for breakfast and supper." It wasn't quite like *that*, but the reality was harsh. Huntsville was an old and failed cotton town. The illiterate outnumbered readers. A statue of Johnny Reb stood sentry outside the courthouse, and racial apartheid underlay every aspect of civic life.

In time, the Germans would help rouse the city from its cultural slumber and backward ways. Their first priority, however, was to help organize the army's new Ordnance Guided Missile Center. Their early work there saw continuing improvements to the motors and guidance systems of the V-2, and created in the process plans for a bigger, more powerful ballistic missile, America's first. It would take the name of the sprawling arsenal around them: Redstone.

Von Braun was landing regular gigs as a public speaker on space travel by now, and his aspirations for grand multistage rockets and space stations had attracted a smattering of press attention—capped by a lengthy April 1951 profile in the *New Yorker* in which he called a journey to the moon "unquestionably a possibility," requiring only "adequate funds and continuity of effort." But it wasn't until that November that he stumbled onto the path to a truly vast audience. Early in the month, von Braun traveled to San Antonio, where the air force was sponsoring a conference on the medical and physical challenges of travel in the upper atmosphere. He was not among the speakers and could stay for just two of the gathering's four days. No matter: "Leaving one of the sessions and stepping to the bar of the hotel," he'd later recall, "I made the acquaintance of a good-looking Irishman who, gazing at the crystal highball between his hands, was sunk in a brown study. 'They've sent me down here to find out what serious scientists think about the possibilities of flight into outer space,' he growled. 'But I don't know what these people are talking about. All I could find out so far is that lots of people get up there to the rostrum and cover a blackboard with mysterious signs.'"

The Irishman was Cornelius Ryan, an associate editor for *Collier's*, a weekly newsmagazine with more than three million subscribers; he'd later gain lasting fame as the author of *The Longest Day*, a classic account of World War II's Normandy invasion. At the moment, though, he was in over his head on a story that he had a hard time taking seriously. Von Braun offered to help. That night, he joined Ryan for dinner and drinks—quite a few drinks—along with Fred Whipple, the chair of Harvard's Department of Astronomy, and Joseph Kaplan, an expert on the upper atmosphere and a professor of physics at the University of California, Los Angeles. The three tag-teamed the writer, selling the practical possibilities of space travel into the night.

"Whether or not he was truly skeptical, we persevered," Whipple wrote later. "Von Braun, not only a prophetic engineer and top-notch administrator, was also certainly one of the best salesmen of the twentieth century. Additionally, Kaplan carried the aura of wisdom and the expertise of the archetypal learned professor, while I had learned by then to sound very convincing.

"The three of us worked hard at proselytizing Ryan, and finally, by midnight, he was sold on the space program."

The result was a landmark series of stories in *Collier's*, beginning the following March: a "symposium" in print, as the magazine put it, that told "the story of the inevitability of man's conquest of space." An editorial introducing the first installment framed that conquest as an issue of Cold War survival: "What you will read here is not science fiction. It is serious fact. Moreover, it is an urgent warning that the U.S. must immediately embark on a long-range development program to secure for the West 'space superiority.' If we do not, somebody else will. That somebody else very probably would be the Soviet Union." Chosen to write the centerpiece story, Von Braun struck the same tone, arguing for the creation of a giant, wheel-shaped space station to serve as a battleship in low Earth orbit. It would take ten years and a lot of money, he acknowledged, but with it, the United States could preserve peace; in others' hands, such a weapon might make slaves of the free world.

Conventional wisdom holds that the space race began more than five years later, when the Soviets launched the first artificial satellite, *Sputnik 1*. One could just as easily argue that the starting gun was fired in that March 22, 1952, issue of *Collier's*. An explosion of national TV and newspaper coverage greeted the stories, especially Von Braun's piece. Time and again, he was hustled on-camera where, telegenic and smooth, with a precise, nasal clip to his speech, he explained the complexities of space travel in terms that TV viewers unschooled in celestial physics could grasp. Moreover, he sold it as *doable*.

Just like that, fantasy became the future, and Wernher von Braun became its face.

6

TO READ THE *COLLIER'S* SERIES TODAY IS TO RECOGNIZE EARLY forms of the familiar. Von Braun argued, for instance, that the technology needed to build his orbiting weapons platform already existed. Rockets to get the pieces up there simply needed refinement—after all, America had reached space with the two-stage Bumper. "By putting a two-stage rocket on another, still larger, booster, we get a three-stage rocket," he wrote. "The three-stage rocket may be considered as a rocket with three sets of motors; after the first set has given its utmost, and has expired, it is jettisoned—and so is the second set, in its turn. The third stage, or nose, of the rocket continues on its way, relieved of all that excess weight." That pretty much describes the Saturn V.

Von Braun's story in the magazine's second installment, cowritten with Whipple, opened with no shortage of confidence: "Here is how we shall go to the moon." It laid out a plan to ferry parts and equipment into low Earth orbit, where a small army of astronauts would assemble them into three tremendous rockets. Two would be gassed up for a round-trip; the third would be a one-way space truck, lugging the provisions and gear needed to sustain fifty crewmen for six weeks on the lunar surface.

This model for celestial travel—of assembling big rockets in weightless space, thereby obviating the need to muscle stupendously heavy spacecraft off the ground and through the earth's atmosphere—would come to be called Earth orbit rendezvous. It remained von Braun's preferred approach for a lunar mission into the 1960s, and it's still considered viable. As you'll see, the notion of sending a space truck to the moon would stick around, too.

Back to the story. "First, where shall we land?" the authors asked. "We may have a wide choice, once we have had a close look at the moon. We'll get that look on a preliminary survey flight." This was an important detail, because scientific knowledge of the moon's surface circa 1952 was restricted to what could be seen by telescope. Which wasn't much.

"Our scientists want to see as many kinds of lunar features as possible," the article continued, "so we'll pick a spot of particular interest to them." Another key point: von Braun viewed a moon voyage as more than a romantic adventure. He'd noted the scientific value of such exploration since at least 1945, and in *Collier's*, it figured prominently.

At story's end, the three ships arrived at the moon and touched down in a tight triangle. "The whirring of machinery dies away," it read. "There is absolute silence. We have reached the moon. Now we shall explore it."

That ushered the third installment, for which von Braun again teamed with Whipple. They wasted little time getting to the action: "The first equipment brought out of the cargo ship is one of our three surface vehicles, tanklike cars equipped with caterpillar treads for mobility over the moon's rough surface," they wrote. "The pressurized, cylindrical cabins hold seven men, two-way radio equipment, radar for measuring distances and depths, and a 12-hour supply of oxygen, food, water, and fuel. Power is supplied by an enclosed turbine driven by a combination of hydrogen peroxide and fuel oil (oxygen escaping from the hydrogen peroxide enables the fuel oil to ignite). The vehicle goes 25 miles an hour on flat ground.

"As soon as the moon car has been set down and checked, a search party boards it to scout out a suitable crevice for the campsite. They drive off in a spray of dust which settles almost immediately, like the bow wave of a motorboat (there is no air to hold the dust suspended, as on earth)." Much of this is important stuff: the front-and-center role von Braun gave the "moon car"; the outlines of its design, which presage the thinking of a decade later; the authors' thoughts on the behavior of lunar dust. By the way, it figures that hydrogen peroxide was involved. The V-2 had used it.

Perhaps the most impressive single sentence concerned the scientific aspects of the mission, its "principal aim"—"Our investigations will help us unravel the secret of the universe: how the moons and planets were born and what they're made of." This might have seemed an overreach to some *Collier's* subscribers, but it would prove to be prescient.

Finally, there was this: "Back [on Earth], a special panel of scientists remains in constant session, as it will all during our six-week stay. A dozen specialists in fields like astronomy, astrophysics, geophysics, minerology, and geology follow our every move by radio . . . keeping track of our findings, suggesting new leads, and occasionally asking for the repetition of an experiment." That describes what would be known as the scientific "back room" at NASA's Mission Control Center in Houston. With a single exception, the experts there wouldn't get to speak to the astronauts directly, but their suggestions would reach the moon, just as von Braun and Whipple foresaw.

Naturally, some elements of the series are off the mark. The authors had the explorers bivouacking in a deep crack to avoid incoming meteoroids and subsisting on "canned and frozen food." They deemed TV transmissions from the moon "impractical." They sent their astronauts on a five-hundred-mile expedition in their moon cars, which attests to how cheap life was viewed in the good ol' days. Von Braun's wheel of a space station made it into *2001: A Space Odyssey,* but not into space; our real-life orbital outposts, comparatively messy jumbles of seemingly mismatched parts, have failed to dominate the world.

Then again, this was written when the only comfortable way to Europe was by steamship. The first transistor radio had yet to hit the market. Jonas Salk was still working on his polio vaccine. To find such farsighted, serious discussion by top experts in a respected weekly was nothing short of astounding to the readers of 1952.

If *Collier's* launched von Braun toward celebrity, Walt Disney, of all people, made him a household name. Eighteen months after that third magazine installment, Disney was planning his namesake amusement park on 160 acres of cleared orange grove in the suburbs of Los Angeles. He'd divided the park into themed "lands" tied in to some of his best-loved movies and characters: Fantasyland was anchored by Sleeping Beauty's Castle, for example, and Frontierland was an immersive ad for TV's *Davy Crockett.* But the animator turned impresario was stumped coming up with a thematic focus, a familiar story, for Tomorrowland.

Until, that is, one of his animators showed him the *Collier's* series. Excited by both its scientific surety and optimism, Disney recognized that here might lie his park's missing ingredient, as well as the chance for some compelling TV. He had his people seek out Willy Ley—von Braun's old colleague from his Berlin days, who had fled Germany in 1935 and established himself as a leading American science writer. Ley had helped Cornelius Ryan polish von Braun's magazine pieces. Now he urged the Disney organization to partner with the rocket engineer.

And so, on two memorable nights in 1955, American TV viewers were treated to episodes of ABC's hourlong *Disneyland* program that, via animation and live-action vignettes, brought to life von Braun's vision of space exploration and introduced millions to the man himself. The first, "Man in Space," aired on March 9. When the camera found von Braun, he was in shirtsleeves, seemingly caught in the midst of an important engineering problem. He stepped away from the task to brief his audience on the workings of a spaceship, using a silver, three-foot model to explain its four stages. He was built like a linebacker, and the voice paired with his bulk was discordantly high pitched. All the same, when he uttered his closing line in that oh-so-precise German accent—"If we were to start today on an organized, well-supported space program, I believe a practical passenger rocket could be built and tested within ten years"—millions believed.

They had to wait more than nine months for the second episode, and a fuller dose of von Braun. For "Man and the Moon" he wore a double-breasted suit, a pale orange tie, and a deep tan. Using a slide rule as a pointer, he walked viewers through a launch, Earth orbit rendezvous, and lunar flight, lacing his speech with specifics. A rocket would be fifty-three feet long; it would top out at 18,468 miles per hour; his space station would orbit exactly 1,075 miles above the earth. Viewers were sold. By hour's end, von Braun was emergent as the man who would get America into space.

He and Willy Ley left their fingerprints all over Disney's newly built Tomorrowland, too. Its chief attraction was the TWA Moonliner, a stylish seventy-six-foot rocket in which visitors embarked on a simulated ride to the moon that drew from the TV shows. At the nearby Space Station X-1, ticket holders stepped aboard von Braun's imagined orbiting platform, stripped of its Cold War weaponry and offering a panorama of cities, farms, and woodland from five hundred miles up. As it passed into night, the towns below glowed. By the standards of the midfifties, it was mind-blowing.

At Redstone Arsenal, however, von Braun's reality was far more earthbound. He wasn't building moon rockets or space stations; he and his Peenemunde team were still designing and testing missiles, and that didn't appear likely to change anytime soon. The International Geophysical Year was approaching in 1957–58, during which the United States hoped to launch a scientific satellite into space, but the Germans weren't part of the project. They were sidelined in favor of the navy, which proposed to put a complex device into orbit—far more ambitious than the minimalist satellite von Braun had pitched—aboard an elegant launch vehicle derived from a research rocket, versus the army's overtly military Redstone missile variant.

Plus, there might have been some queasiness within Dwight Eisenhower's administration about von Braun's past. He and much of his team had become American citizens by now. But the navy rocket hadn't been built by ex-Nazis. The navy hadn't bombed London.

7

AT ABOUT THE TIME WERNHER VON BRAUN WAS BECOMING A household name, Mieczyslaw Gregory Bekker was retiring from a first career as a military engineer in Europe and Canada, and beginning a second in the United States in which he, too, would earn a measure of fame. It was a modest prestige, next to the rocketeer's. He didn't seek, or receive, recognition beyond his professional colleagues, and even if he had, his field of study wasn't nearly as romantic as space travel. He lacked von Braun's Hollywood looks, too: vaguely elfin in appearance, Bekker stood five foot three in his shoes, and that was

M. G. "Greg" Bekker. (FACULTY OF AUTOMOTIVE AND CONSTRUCTION MACHINERY ENGINEERING, WARSAW UNIVERSITY OF TECHNOLOGY)

with the help of a lofty head of thick white hair. Another thing: the average American could spell "Wernher."

Lucky for us, he's better remembered as Greg, which is what his Western friends called him, or as M. G. Bekker, the handle he used in his books, journal articles, and professional correspondence. Among a certain segment of transportation scholars, the latter name is preeminent. He almost single-handedly created the study of off-road mobility, and he named it—terramechanics, which explores, more specifically, the relationship between off-road vehicles and the soils on which they travel. Engineers around the world refer to the Bekker method or the

Bekker model. And anyone who's ever gone mudding in a Jeep, or raced motocross, or bulldozed, or driven a tank—who has relied on knobby tires or caterpillar tracks to cross mud or sand, bog or ash—owes something of that experience to the second member of our cast.

Bekker's early life is not nearly so well documented as von Braun's. Seven years older, he was born in May 1905 in the Southeastern Polish village of Strzyzow. He attended the Warsaw University of Technology, held an internship at a Renault automobile factory outside Paris, and graduated in 1929. He married Jadwiga Rychlewski, who worked in a Warsaw flower shop. After serving two years in uniform, Bekker returned to the university to teach in its military school and start its Special Vehicles Laboratory. Its assignment: evaluating off-road cars and trucks offered to the armed forces, and devising improvements to tracked vehicles.

This was pioneering duty. In the twenties and early thirties, horses still played central roles in army life, and for good reason—and that reason was *mud*. Throughout human history, this near-ubiquitous nuisance had been one of the greatest impediments to travel, commerce, agriculture, and prosperity, let alone fighting wars. And where mud wasn't, there was likely loose sand to contend with, or snow, or swamp. The automobile, still in its skinny-tired adolescence, was at a distinct disadvantage: Its wheels might be an efficient means of transporting loads over smooth, hard surfaces, but such surfaces were hard to come by. Vast expanses of the world were roadless, and, in most places, even established routes went unpaved. Often as not, they consisted of rutted wagon tracks that devolved into gluey wallows after a rain.

This was a problem in America, as elsewhere. Modern road building had been under way in the United States for just a few years when Bekker went to work, and progress proved slow. It wouldn't be until the late 1970s that our paved public roads finally outdistanced the unpaved; even today, one in three miles of publicly maintained American road is gravel, clay, or of some similarly rude construction. That said, Poland and the rest of eastern Europe had more than their share of rough country. Figuring out how to cross it became Bekker's preoccupation.

He studied what had been tried before, and why so little had worked. Tanks were a new technology, having just debuted in battle toward the end of World War I. But Bekker found that what gave them so much promise as weapons—their caterpillar tracks—was an idea that dated back to 1770, when an English patent for a carriage used slats of wood for its tread. That invention had gone nowhere, literally, and the same went for a host of derivations that appeared throughout the nineteenth century. Part of the problem was that a caterpillar track is subject to a lot of stresses, which early designers addressed with the durable materials available to them—namely, iron and lumber. That produced vehicles so heavy that any advantage offered by their tank treads was offset by their propensity to sink. And in those horse-drawn days, it took a mighty strong team to pull them.

An American farm tractor, the Holt, finally married lighter steel and engine power, and became the basis for tank design during the Great War. Since then, U.S. agriculture had led the way for off-road machinery worldwide, and engineers hit upon a rule of thumb: the lighter a vehicle's "ground pressure," the better it usually performed. Ground pressure was determined by dividing a vehicle's weight by the area of its contact with the ground. The heavier the vehicle, the bigger the contact area had to be. Meaning wider, bigger tires, or wider and longer tank treads.

But not always. Sometimes greater ground pressure yielded better traction. Sometimes small studs on a tire grabbed better than big ones. No one understood why, which was reflected in the army hardware Bekker evaluated. In Poland, as everywhere, many off-road designs seemed the product of educated guesswork; even the successes appeared as much the product of luck as science. It occurred to Bekker that the interface between a vehicle and the ground had to be governed by physical laws as sturdy and sure as those governing the relationship between a ship's hull and water, or an aircraft's wing and the air it rode, and that it should be possible to use those principles in vehicle design. Theoretically, there should be formulas that explained it all: an engineer should be able to work out a viable design with calculations, rather than rely on slow, expensive trial and error. Finding those principles and formulas was on Bekker's mind when Hitler's Germany and Joseph Stalin's Soviet Union—wary allies at the time—invaded Poland in September 1939.

Bekker, his wife, and their two small daughters retreated to Romania, then to France. He had just started with the French Ministry of Armaments when, in June 1940, Paris, too, fell to the Nazis, again forcing the family to flee. The four spent the next two years in a hotel in Hyeres, on the Riviera, Bekker toiling over his formulas in their room. "He had his notebooks," his older daughter, Eva Heuser, recalled, "and he was writing equations in them." In 1943 came an opportunity to move west. Commissioned an officer in the Canadian army, Bekker took charge of the service's research into vehicle mobility.

By that time, he was almost certainly honing the observations, seemingly simple, that would undergird his major contributions to come. First and most important: no one vehicle can perform well everywhere, because soil types and conditions vary so widely that a one-size-fits-all approach cannot succeed. With that in mind, it follows that, second, understanding soil well enough to predict its behavior over a range of climatic conditions is essential to success.

Bekker realized that evolution in the animal kingdom illuminated these points. The first creatures to move around on land were amphibians living in bogs so soupy that they supported themselves on their bellies, and used their legs more to paddle than to walk. Over eons, much of the wetlands firmed up, and "walking, running, and even jumping became possible," he wrote later. "This led to a new shape of extremities and to a definite elongation of those animal limbs that served the locomotive purposes.

"In the Cenozoic period, the abundance of flat, relatively strong ground produced further changes in structure and the further elongation of legs," he theorized. "In general, it must be said that the strength and deformation characteristics of various soils have played a definite role in the . . . form, size, and weight of the animal body."

This view of adaptation to the terrain offered an example for engineers to follow. The key was understanding the precise nature of what lay underfoot—the properties of cohesive soils, such as clay, versus those of loose sand, soft mud, and the muskeg that covered much of the Arctic—and coming up with formulas to describe those properties. Much of Bekker's new home was bound in snow and ice for half the year or more, so deciphering how snow clumped and compressed, its tendency to bind or slip, occupied a good piece of his time. He examined ski and sled design, traveled to Switzerland to sample alpine snow, and consulted with engineering faculty at universities all over the world, especially in the United States. He spent long stretches at New Jersey's Stevens Institute of Technology.

The immediate payoffs were some novel vehicle designs. One, a trooper carrier for duty in the high latitudes, looked more like a spacecraft than a troop transport: low slung and fast, the Cobra, as Bekker dubbed it, was articulated—cut in two, with its front and rear halves each propelled by tracks. The halves were linked by a joint that permitted some degree of independent movement, the better to help it maneuver as it sped along. He invented a new track design for the Cobra, too.

By the midfifties, Bekker was more or less on loan to the U.S. Army and received an offer he found irresistible: a chance to lead a newly created mobility research program within the army's Tank-Automotive Command at the Detroit Arsenal, with the budget and freedom to organize and staff it as he saw fit. He retired from the Canadian service a lieutenant colonel to become the civilian director of the Land Locomotion Research Laboratory.

At the same time, Bekker was finishing his first book, *Theory of Land*

Locomotion: The Mechanics of Vehicle Mobility. Published by the University of Michigan Press in 1956, it remains the foundational text for the study of vehicle-soil mechanics. Its bottom-line message—know your terrain in extreme detail, and adapt your vehicle's form, size, weight, and power to suit it—was made in page after page of formulas that I would not attempt to explain here, even if I could. Suffice to say that it introduced the Bekker method, an integrated theory of overland travel, theoretically enabling engineers to assess an environment and mate it on paper with a vehicle that should work there. The book was instantly celebrated as vital and overdue. The Bekker method is still talked about and debated over, which is a pretty decent measure of its importance. The literature is filled with scholarly complaints about its limitations, but, over the decades, no one has managed to fully replace it.

Once established in Detroit, studying mud mobility, Bekker realized he faced a challenge equally as sticky: he couldn't find qualified engineers to work with him. "It is almost impossible," as he told listeners at a government mobility symposium, "to find personnel with interests and educational background which could properly aid the research."

He would soon find an exception. The right man, however, was thousands of miles away, and he had his hands full.

8

ONE SUNNY, COOL MORNING IN JUNE 2019, I DROVE UP A SUCCESsion of steep, winding roads into the hills high above Santa Barbara, California, to a spectacular modernist house all but invisible from the curb behind a screen of citrus trees. It was long, low, and fashioned of stucco and dark wood; out back, floor-to-ceiling glass offered a panorama of the white-flecked Santa Barbara Channel, the rugged Channel Islands, and the wide Pacific.

The man who came to the door wasn't what I expected. I knew

Ferenc Pavlics to be ninety-one, and this fellow did not fit the model. He was tall and sturdy, with a shock of wavy, white hair and a face remarkably free of lines and sags. He led the way through the house with a loose-jointed, athletic grace and pointed me to a sofa under the living room's sloped ceiling, opposite a massive fireplace of native stone. As I pulled a notebook from my backpack, he held up a hand in warning. "My memory's fading," he said, taking a seat himself. "My thinking abilities are declining. I'm getting old."

"You seem to be doing all right," I replied.

"Well," he sighed, "it may not be obvious to you. But it's obvious to me."

I spent two days with him, during which, near as I could tell, his memory was sharp. He told me the story of his professional life in detail. He answered my questions without hesitation. He explained complex engineering problems, and their solutions, so that I understood.

Introducing the third member of our cast: Ferenc Pavlics, "Frank" to American friends and colleagues. Born in February 1928 in Balozsameggyes, a village in western Hungary—then, as now, an outpost of small, tile-roofed houses bunched around a rural crossroads ringed by farmland, and miles from its nearest neighbor. Pavlics was the fourth child of nine born to a pair of teachers; his mother taught the lower grades in the village school, and his father, the upper. He was a good student, and from an early age demonstrated a "love to do things with my hands and to build things," he told me. "I always have been mechanically inclined."

He rode a train twenty miles to the nearest city, Szombathely, for high school, where he initially felt a pull toward chemistry. An experiment gone awry in the family's kitchen—"my little sister got the experiment all over her dress"—encouraged his shift to mechanical engineering. He sought, and won, admission to the Technical University of Budapest.

Upon his graduation, in 1950, Pavlics landed a job as an engineer at a government-run industrial design institute in Budapest. There he developed tools and equipment for factories, calculated how to best arrange the pieces of assembly lines, built new plants, and renovated old ones. He moonlighted in the evenings as an assistant professor at his alma mater. And on the rare occasions he wasn't working, he joined his friends rowing on the Danube. "The boats had two rowing seats with paddles, and a seat on the back for steering it," Pavlics said, Hungary still strong in his speech seventy years later. "And we always sit a pretty girl on the front, to look at."

Which is how he met Klara Schwab, an accountant two years his junior. "She started out to be the pretty girl," he said. "She turned out to be much more than that." At a boathouse bar one evening, "a friend of mine wanted to get with her and try to make a date. She said, 'No, but why don't you send over your friend?'" She meant Pavlics, who was tall, fit, and both quick minded and humble. They married not long after.

It was a good life, by the standards of the Eastern bloc. The couple had an apartment, engaging jobs, and got around the city easily by bus or bike. Six years after starting at the institute, Pavlics's future was full of promise. Yet he felt a growing discontent: the Soviet lock on Hungarian life was complete and stultifying; their puppets ran the country, their troops were quartered there, and Pavlics and his colleagues had no contact with their counterparts in the West, no access to the literature, no exposure to the "state of the art" elsewhere. Hungary's scientific and technical communities were straitjacketed.

In October 1956 he learned that his frustration was shared by most of the population, and watched it come to a head. University students in Budapest marched to the Parliament Building with demands for free elections, evicting the Soviets, and Hungary's economic make-over. When they came under fire, the protestors didn't back down. In hours, the demonstration spawned a revolution, and it spread fast; Hungary's own army joined the uprising, a new government started to organize, and the Soviet military cleared out of Budapest. Hungarians rejoiced.

"We were all excited about the thing," Pavlics remembered. "It was a Russian occupation for a long time, and we hated the Russians. So it was a big relief when the revolution appeared to be succeeding."

He was not among the fighters. His revolutionary role was that of a management organizer; his Communist supervisors took off, leaving the institute without leadership, and he stepped in. But Pavlics was elated by the Soviet retreat and the prospect of a free society, right up to the moment everything reversed. "It took about ten days for the Russians to send in new troops and beat down the revolution," he said. "It was very bloody."

Within the city, the revolt was snuffed out in a matter of days. The Soviets quickly moved beyond Budapest to consolidate their control of the countryside. By late November, Pavlics decided he and Klara had to get out, and be quick about it. They left Budapest by train, carrying little in the way of luggage, and stepped off at a station twenty miles from the Austrian border, the edge of a zone requiring special papers for travel. From there they struck out on foot for his sister's house on the heavily fortified frontier. "We had to walk that twenty miles," he said. "Some of my friends were caught on a train going in a western direction, taken off the train, and put on another train back to Budapest. So we walked the twenty miles, mostly at night."

They avoided roads, cut wide around villages, sought the cover of trees, slept when and where they could. It took three days to reach his sister's place. Their plan for the last, most dangerous leg of the journey was sketchy. His brother-in-law was a doctor on good terms with his neighbors, among them a butcher who delivered meat to troops guarding the border. "He made an arrangement with the butcher to take us with him," Pavlics said. His brother-in-law gave him his own identification papers, and his sister gave Klara hers. If they were stopped, the doctor told them, they were to say that Pavlics was calling on a patient.

Accompanied by a sympathetic Hungarian soldier, the party left the house on a horse-drawn wagon loaded with meat. Before long, they came to a Soviet roadblock. The officer in charge demanded their papers. The Pavlicses handed them over, sure the jig was up: neither looked at all like their in-laws; a glance at the mugshots affixed to the documents would be the end of them. Except that the officer was training a group of new arrivals, showing them how to check papers for the necessary stamps, dates, and signatures—and never paused his lesson long enough to take a good look at the couple before him. He waved them on.

It was raining and cold as they reached the border, a wide and muddy no-man's-land scraped bare to afford clear shots at would-be defectors. Mines lay in wait. Troops and attack dogs patrolled. But there were corridors through the mines, the friendly soldier explained, and several minutes passed between the patrols. He'd give them a sign when the time was right.

And so, on the night of November 29, 1956, Ferenc and Klara Pavlics joined nearly two hundred thousand of their countrymen fleeing Hungary during and after the uprising. Penniless, their only possessions what they could fit in a briefcase and an overnight bag, they staggered and slopped across the strip toward the lights of a village in Austria.

Bused the next day to a sprawling refugee camp, they were "processed" by Western authorities and asked where they'd like to resettle. "I had no idea what I was going to do," Pavlics recalled. "I didn't speak English at all. I spoke a little German." It made sense to apply for a West German visa. A few days later, a representative of the country appeared in camp to interview him, and by session's end, he had a job offer in the Baltic port city of Bremerhaven. The visitor promised to take care of the paperwork.

Weeks passed in the camp with no further word. Fretful, the Pavlicses decided they needed a backup plan, so they traveled to the U.S. embassy in Vienna and joined a line of refugees that stretched around the building. They finally reached the doors the next day and applied for American visas. Go back to the camp, they were told. We'll be in touch.

Ferenc "Frank" Pavlics,
early in his career at GM.
(PAVLICS FAMILY)

Not long after, they learned that they'd been granted entry by both countries. "So, we had a decision to make," Pavlics said. He favored Germany, but Klara argued against it. "I don't want to stay in Europe," she told him. "Trouble always starts here, and I'm through with wars. Let's get out of here." Pavlics broached a compromise. The U.S. government had arranged for them to travel by rail to Bremerhaven, where they would board a ship for New York. So let's take the train, he suggested, and if we like the look of Bremerhaven and think we can be happy there, we'll stay put. If not, we'll get on the ship.

They arrived on the Baltic coast in the first days of February 1957. Bremerhaven was cold, gritty, and gray—"an ugly, industrial city," as Pavlics remembered it. He was as eager to leave as his wife was. They made their escape on the U.S. Navy troop ship *General Nelson M. Walker*, which was stuffed with four thousand other Hungarians. Men and women were separated to crowd into airless berthing compartments. They slept on bunks stacked four high.

Nine days, it took the ship to cross the North Atlantic. "A lot of seasick people throwing up on the deck," Pavlics said. "It was not a pleasant trip, with the people throwing up on the deck around you, and the heavy seas. To avoid seasickness, I spent most of my time on deck, to get fresh air. Of course, in February's stormy weather, the saltwater waves were splashing up over the deck, so that when we arrived in New York, my winter coat—which was a long, heavy winter coat—it could stand up by itself because of the salt."

In New York, they were herded into buses and carried twenty-odd miles to Camp Kilmer, a dormant New Jersey army base that had been reactivated to receive them. The Eisenhower administration was enthusiastic in its welcome: it housed, clothed, fed, and resettled more than thirty thousand refugees at Kilmer in the first half of 1957, and encouraged employers throughout the country to tap the pool of new talent. "We at the time were very appreciated by the West," Pavlics said, "because we were freedom fighters."

Five days into their stay, Ferenc and Klara were approached by a short man with a head of thick, white hair who, speaking in German, told them he was looking for engineers. He said he ran a lab in Detroit, where he studied the science of vehicle mobility. After they'd talked for a while, he invited Pavlics to join him.

The new arrival had nothing in writing that proved he was an engineer. Neither did a handful of other Hungarians the visitor hired that day. Years later, Pavlics asked Greg Bekker why he'd taken such a chance on him, why he'd taken the word of a stranger.

"Well," Bekker replied, "if you didn't work out, we'd have just fired you."

9

I WILL PAUSE THE ACTION HERE TO CALL ATTENTION TO ONE DE-tail of our story so far, on the unlikely chance that it has slipped past without your noticing. And that is that the three characters we've met, all destined to play key roles in the Apollo program, were foreign born. And they were by no means outliers. America's race to reach the moon, both within NASA and at the aerospace companies that built the hardware, relied on the minds and talents of immigrants—on Americans who happened to start their lives elsewhere.

One of them now departed Camp Kilmer for Detroit, leaving his wife behind until he'd earned his first paycheck. "Hungarians kept coming, and they were running out of space, so they wanted to get rid of people as soon as possible," Pavlics told me. "They gave me a train ticket and five bucks, and said, 'You're on your own.'"

On the platform in Detroit, he encountered a Polish couple holding a sign bearing his name. Part of a local Catholic charity, they put him up, drove him to work and back, and fed him for several weeks; the household communicated in sign language and mime. At the Land Locomotion Research Laboratory, Bekker assigned Pavlics work as a draughtsman. It was an entry-level post.

Like his fellow Hungarian hires, he was eager to make the most of this new opportunity. They all took English classes at Wayne State University. "It's amazing how fast can you learn a new language when you must," he reflected. "In about three months, I could get along, shopping and interfacing with colleagues at work. My boss spoke German, so we were communicating in German in the beginning, and then as I learned English, we communicated in that language." Once Klara joined him, the Pavlicses moved into a small apartment. A colleague gave Frank lifts to the lab and back until he learned to drive and bought his first car. "I had an old, used Plymouth," he said. "My friends and I went out on big parking lots where there was no business, and practiced driving."

After a few months, Bekker promoted Pavlics from draughtsman to engineer. The lab's chief task was to investigate ways to improve the mobility of tanks and other army vehicles. Pavlics's role was to design equipment to test the lab's theoretical work—which, thanks to Bekker, was without peer. They formed a close partnership, founded in part on a shared gentlemanly, Old World reserve, on "having that bit of formality that seemed more characteristic of European immigrants of that era," as Pavlics's son Peter put it, "where they were never quite comfortable with the immediate familiarity of Americans.

"I have never seen my dad in a T-shirt—and in shorts only on the tennis court," he said. "My guess is that Greg was that way, probably even more so. Classy, in the best meaning of the word."

Pavlics was a few months into the job when, on October 4, 1957, the world was stunned by the Soviet Union's launch of its *Sputnik 1* satellite. Wernher von Braun's team at Redstone—his outfit by now reorganized and renamed the Army Ballistic Missile Agency, with von Braun as its technical director—looked on as the navy rushed a catch-up satellite launch two months later. But its Vanguard rocket blew up on the pad, to the administration's deep embarrassment, and with that, von Braun got the green light to take over. His team put the first U.S. satellite, *Explorer 1*, into orbit in January 1958, aboard a four-stage Juno I, a derivative of the Redstone missile. Had he been given the job to begin with, there's a strong chance that he'd have beaten the Soviets into space by more than a year.

The Soviet Union's second spaceflight put a dog in orbit, and word reached U.S. intelligence that a man was next. America raced to do the same. In October 1958 Congress created NASA to manage the country's nonmilitary pursuits in space. A month later, the new agency organized the Space Task Group, a think tank of engineers at Virginia's Langley Research Center that would later evolve into the Manned Spacecraft Center in Houston. Three weeks after that, the task group had Project Mercury up and running.

This flurry of interest in space made an impression on Greg Bekker. "I had much freedom in pursuing Soil-Vehicle inquiries," he recalled in a letter years later. "In 1957 I became intrigued with the problem of surface locomotion on the moon." At the time, Pavlics was designing and building soil bins, which are to ground mobility what wind tunnels are to aeronautics. The bins could be filled with different types of soils, and their relationship with wheels and tracks of varying design measured and analyzed. The bins themselves tended to be a few feet wide and deep, and up to one hundred feet long. Some were circular, to create a continuous bed on which wheels could be tested without stopping for hours straight. Machinery positioned above or alongside the bins rolled a wheel or track over the soil at preset ground pressures, speeds, and torque levels. Instruments measured their interaction.

Bekker occasionally repurposed the bins, as well as the lab's other equipment, to experiment with finely ground pumice, which he reckoned would simulate the character of lunar soils. That was a guess, he acknowledged later, as "even such simple question as the effect of low gravity and vacuum upon the intergranular and molecular forces of soil mass, was totally unknown." It was no sure bet he would ever put his insights to use, either. The country had not publicly declared its intent to put astronauts on the moon—in fact, it had no astronauts at all until April 1959, when NASA introduced its "Mercury Seven" to the world. The new space agency was still years away from strapping a human into a rocket for even suborbital flight. But with *Sputnik* and *Explorer* and the possibilities of space a new national obsession, and a race under way with the Soviets, it seemed plain that lunar voyaging was only a matter of time. And once on the moon's surface, American explorers would not be content to walk everywhere. They'd want to drive.

10

THIS WAS NOT A NEW IDEA. PEOPLE HAD BEEN THINKING ABOUT driving on the moon for most of the century, though the bulk of that thought was embedded in science-fiction stories. One fictional description of a pressurized rover cruising the lunar surface dates to 1901. Others came along regularly in the decades after: rolling spheres, cigar-shaped cruisers, walkers and crawlers, tanklike transporters. The *Collier's* series had been among the few nonfiction entries on the subject, assuming you can call such futurism nonfiction. Now another appeared—a most unexpected addition to the roster of early rover concepts, from a most unexpected source.

The machine envisioned by von Braun's old mentor, Hermann Oberth, was so fantastical that it is often dismissed as whimsy. Oberth did not see it that way. He offered up his book *The Moon Car* without the slightest hint of tongue in cheek; he suggested as much with the title of his first chapter, "Principal Considerations," which is about as no-nonsense as you get. His dedication—"To the great engineer and organizer Professor Wernher von Braun, who transformed the idea of space travel into reality"—was sincere as well.

Oberth had been nursing his concept since 1953. He'd reviewed the ideas of others, including von Braun, and concluded they fell short in a key respect. "What are the explorers going to do if they encounter a deep chasm on the lunar surface?" he asked himself. "Such chasms would force the explorers into long detours and ruin the schedule for the exploration; and chasms of such type are apt to be numerous on the moon."

"In that manner, my moon car originated," he explained. "It can ride and fly, or better, it can ride and jump. In fact, it can make very considerable jumps which should increase the range and improve the visibility of the driver."

Oberth included a brief exposition on his moon car in a 1954 book, *Man into Space: New Projects for Rockets and Space Travel*, but he continued to build on the idea—and evidently did so in Huntsville, because the following year, he accepted his protégé's offer of a post at the Redstone Arsenal and stayed until his return to Germany in 1958. His treatise on his riding and jumping rover, translated from the German by Willy Ley, saw print on both sides of the Atlantic in 1959.

And, oh boy. Oberth's moon car was shaped like an overgrown Tootsie Pop, with its spherical cabin up top of a single long leg, which in turn was mounted on a caterpillar-tread foot. The leg fit inside a cylinder that passed through the middle of the cabin and, with compressed gas, could be drawn into or pushed from this sleeve. For most of its travels, the moon car would cruise around with the leg withdrawn, so that the whole contraption measured a petite 40 feet tall, maintaining its top-heavy balance with a gyroscope on its roof. When necessary, it could extend the leg by 20 feet with a blast of gas, causing the whole vehicle to leap as high as 410 feet. "The external shape of the vehicle is . . . uninfluenced by considerations of air resistance," Oberth pointed out. "It could have any weird shape."

Studded with detailed drawings and equations, Oberth's book addressed technical questions about the moon car's inner workings ranging from the gyroscope's design to the location of the bathroom. With a lunar weight of 3,640 pounds and a solar array generating seventy horsepower, Oberth wrote, the moon car would top out at about ninety-three miles per hour. That sort of talk didn't do much for his credibility. Neither did the book's frontispiece, a cartoonish illustration of the moon car zooming up a steep mountainside, a spaceman dangling on a rope ladder unfurled from its hatch. Granted, he'd always been a theorist more than a hands-on engineer. And sure, he'd grown into a bit of a crank in his middle age. But it was hard then, and it's even harder now, to read *The Moon Car* without suspecting it to be an elaborate prank.

The book was not without its contributions, however, and one was its opening paragraph:

> When the first expeditions land in the lunar deserts, the men will need usable vehicles. I have been thinking about such vehicles for a long time, and it is no longer "too early" to publish my thoughts. True, another decade or two may go by until the first men set foot on the moon. But when this happens, it would be practical if such a vehicle had already been developed and (as far as this is possible on earth) had been tested, too. It will also be practical not to hurry such a development

at the last moment because a moon car must function under conditions very different from those that apply to our own cars, and it must be based on fundamentally different principles.

If you remember nothing else about Hermann Oberth's ideas on the subject, remember this: when developing a moon car, it's best not to rush things.

11

BACK AT THE DETROIT ARSENAL, GREG BEKKER FOUND HIMSELF barred from further inquiry into lunar mobility. "Unfortunately, the army has found that they have no mission and budget for lunar soil explorations, even from locomotion viewpoint," he wrote later, "and experimental work stopped." But the Land Locomotion Research Laboratory took up other research, wholly terrestrial in nature, that would prove helpful to lunar studies in the not-distant future. Frank Pavlics, now chief of the lab's Experimental Design Section, was overseeing the construction of scale models for use in the soil bins. They enabled him to study the behavior of entire vehicles in different types of soil and terrain, rather than just wheels or tracks. Model building would become key to his future inquiries.

As it happened, they didn't have to wait long to get back into their lunar experiments, because in late 1960 Bekker received another sweetheart offer. General Motors had recently created a new Defense Systems Division, with offices in Detroit. Its programs catered to the varied needs of the U.S. military. One would handle sea systems, with the navy as its prime collaborator and customer; another was to work with the air force; and the third would focus on off-road mobility research. How would Bekker like to tackle that last assignment, as its top scientist? Bekker had one demand: "My joining the GM was conditioned by me to the effect that Frank be hired too. He is a fine man and excellent engineer," he wrote later. "I was fortunate that my 'right hand' . . . went with me to General Motors."

For Pavlics, the decision was easy. "He was a very, very capable engineer," he said of Bekker, "and he was the best boss I ever had."

The deal got even sweeter shortly after they made the move. "The two of us joined the GM organization in Detroit and set up a new lab for GM," Pavlics told me. Then "General Motors decided to move the lab to Santa Barbara. For the navy support, we needed access to the ocean to do underwater acoustic experiments and all that.

"GM asked everybody, 'Would you accept a transfer from Detroit to Santa Barbara?' They sent my wife and me out here for one week to look around and decide whether to accept the transfer." The Pavlicses found the California city a stark contrast to their Michigan home. Blessed with breezy Mediterranean summers and warm winters, Santa Barbara is wedged onto a narrow alluvial plain, its back to the steep, chaparral-clad Santa Ynez Mountains, the ocean at its feet. Tall palms line the beach, which is on a rare stretch of the West Coast that faces south. It might be the loveliest spot in the state. Frank and Klara weighed the decision for all of ten minutes.

The new research laboratories of GM's Defense Systems Division were housed in a campus of modernist buildings in Goleta, California, just west of the Santa Barbara airport, where avocado orchards were fast giving way to high-tech firms and tract homes. GM bought the property in August 1960; within fifteen months, four hundred people worked there, some hired from within GM, others from the competition, and a few straight out of college.

Bekker and Pavlics labored in a sleek, two-story office building that fronted on Hollister Avenue, then a crooked, two-lane country road. Out back, a minute from their offices, stretched an immense windowless lab and, behind that, another office building for the division's engineers. All told, the complex boasted a quarter million square feet of work space surrounded by acres of open land. Pavlics soon converted part of it to an off-road proving ground.

Their supervisor was a New Jersey native and navy veteran named Sam Romano—smart, tighty wound, and ambitious—who, within the lab's Land Operations Department, headed up its Vehicle Systems Office. A shipboard fire controlman during World War II, Romano discovered he had a knack for troubleshooting radar and sonar equipment. After his hitch, he used the GI Bill to earn a degree in mechanical engineering, then spent six years with the Avion Division of ACF Industries, a defense contractor, developing navigation displays for military aircraft.

Two of Romano's bosses at Avion were hired away by General Motors, and in 1960 they convinced him to join them in Detroit. So as everyone settled into the new Santa Barbara lab, the GM hierarchy directly above Greg Bekker and Frank Pavlics consisted of three former Avion colleagues who happened to share an excitement for aerospace projects. Romano had spent most of his time at Avion as a lab manager, which prepared him well for his role at GM—close to the lab's work, while not enmeshed in its detailed engineering. His job was more that of a project CEO: he managed the budget and schedule, ensured the efficiency of personnel and assets, vetted the ideas emerging from the lab, and tried to keep his projects moving at a reasonably brisk pace.

But his primary mission, shared with everyone in the new complex, was to chase government business. GM had created the division to boost the company's share of defense orders, which had fallen sharply during the 1950s. "They wanted to get involved in off-the-road vehicles—military cross-country vehicles, agricultural vehicles—and we did the research work," Pavlics explained. "We had to be self-supporting, so we had to bid on jobs and get contracts. Only in the very beginning did GM underwrite our costs."

Even so, almost from the day they moved into the new place,

Sam Romano, pictured during his tenure at Avion. He'd soon become the go-getting boss of GM's lunar program. (ROMANO FAMILY)

Bekker and Pavlics were experimenting with rover concepts—which didn't promise a quick return on investment, to say the least—and did so with the corporation's blessing. Not only did Romano and his bosses share Bekker's interest, but the division's head in Detroit, Harold Boyer, might have been an even bigger booster. The GM vice president was intrigued by America's early space ventures and eager to see the company find some role in them. "Boyer wanted to give something back," said Paul Jaquish, an engineer who came to the lab from a GM operation in Saginaw, Michigan, and helped Romano manage its early lunar work. When the lab started, "GM had no particular lunar ambition, and certainly not a roving vehicles ambition. But when it came along, it fit nicely into what Harold Boyer wanted to do."

The Santa Barbara operation would host large personalities and a few real characters over the coming decade, so it's little surprise that its lunar program would be led, in the long term, by such a weird but complementary triumvirate as Romano, Bekker, and Pavlics. "Sam was a very intense guy," said John Calandro, who'd worked with Bekker and Pavlics at the Detroit Arsenal as a college intern and, upon graduating, landed a job with GM just as the lab moved west. "He was a hard-driving kind of guy. He was a salesman." Dark haired, dark eyed, and fast talking, Romano had high expectations and limited patience. "There was very little joking with Sam," Jaquish recalled. "It wasn't that he didn't have a sense of humor, but that was for later—that was for after hours. My experience with Sam was that when you were working, you were *working*."

"He wouldn't take no for an answer. His workers either loved or hated him," Romano's son Cliff told me. "He didn't have a tolerance for crap going wrong. So, if things weren't going the way he wanted them to, he became difficult to deal with."

"He had goals," his wife, Marge, said, "and he would see to it that you stayed on top of a problem, and you would do it until you got it right. He was probably very tough."

Pavlics offered a counterpoint to virtually all of Romano's traits. He was quiet, deliberative, unfailingly polite. "You didn't have to manage Frank," Jaquish said. "You mentioned what had to be done, and he went off and did it." Jerry Skaug, an engineer who would work on drive systems for the lab's lunar vehicles, remembered him as careful and "very technical." When asked a question, Pavlics "wanted to know all of the facts, and he would usually not give you an answer immediately. He'd think it over and make sure he had the right answer before he gave it to you."

Bekker, meanwhile, was "a special case," Jaquish said, "an island unto himself" who spent much of his workday immersed in research and "didn't try to get into the management of the place. He was sort of off to the side in the organizational chart." The older engineer was hands-on when it came to lunar projects, however, which could be a challenge for Romano. "I'm sure they had a lot of disagreements, because it would be Greg's way and only Greg's way," Marge Romano recalled. Agreed Jaquish: "He didn't take instruction well."

As often as not, the professorial Bekker prevailed in their disagreements—but then, he'd invented the field. "He thought what he was doing was extremely important," Calandro offered. "And he was right."

PART THREE

PRINCIPAL CONSIDERATIONS

12

BEKKER'S FOUNDATIONAL RULE OF OFF-ROAD MOBILITY—THAT A vehicle's design should rely on an intimate understanding of the terrain it would be called on to travel—ran into trouble when it came to the moon, because no one understood, intimately or otherwise, what the ground there—the *lurrain*—was like. Scientists disagreed over major features, let alone the details: Were the maria, the dark-colored lunar seas that form the facial features of the man in the moon, beds of solidified lava, as many suggested, the remnants of a once-hot lunar interior? If not, what were they? And what about the craters? Were they all carved out by meteorite impact, or could some of them be volcanic, too?

Into this debate had stepped Thomas Gold, an Austrian-born British astronomer, physicist, and sometimes biologist working at the Royal Greenwich Observatory. In a 1955 article for a British academic journal, Gold argued that the moon had been shaped by powerful erosion, despite its lack of water and wind. The agents of change were, rather, fierce bombardments of radiation and meteorites, along with brutal swings in temperature between lunar night and day; over time these destructive forces smoothed mountains, ground rocks into grain, and ultimately reduced everything on the surface to dust— "presumably very fine dust" that migrated from the highlands to the low and could flow "like a liquid" to slowly, steadily, fill in craters and other low points on the surface. The maria, Gold theorized, were not lava flows, but seas of this settled material. In places, it might be a thousand feet deep.

Though the paper nowhere hinted that this dust was unconsolidated—that is, loose—it promptly raised questions. Didn't this mean that a landing spacecraft could promptly sink out of sight? Or that astronauts would find themselves battling a dry lunar quicksand? Gold viewed that as unlikely. He said as much in a short NASA film a couple of years later, by which time he'd moved to the United States. "I think it is mostly that it is a fine dust, but that it has, within a few inches of the surface, frozen itself together pretty well—so you will not, I think, sink in deeply," he said. "So, it's pretty solid. But on the top, there will be maybe an inch, or a few inches—I can't tell—that will be loose.

"By and large, it will be . . . like landing in the Sahara. Or, perhaps, if there were such a thing, the Sahara covered with a few inches of snow."

Unfortunately, Gold also had a flair for the dramatic, and in some of his public remarks, he didn't entirely discount the possibility of spaceships swallowed or astronauts drowned by dust. As he'd put it later, parts of the moon might resemble the surface of a glacier, on which bridges of seemingly solid snow conceal yawning chasms and might break underfoot. If *he* were on the moon, he'd want to be roped to something, just in case.

Fairly or not, he never lived down his reputation as the guy who predicted that a moon landing would beget a moon sinking. Gold was difficult to ignore—he served on the President's Science Advisory Committee and was an advisor to NASA through most of its first several years of existence—and as the impression calcified that he'd made this prediction, it prompted Frank Pavlics and the GM team in Santa Barbara to experiment with rover designs that might work in loose, fluffy regolith.

Perhaps the strangest model they developed was in the form of twin Archimedean screws. Points aimed forward, turning rapidly in opposite directions, the side-by-side screws didn't make for a fast vehicle, but they eagerly dug the model out of any small avalanches the team buried it under. "We built a soil bin which was about five feet wide, three feet deep, and about fifty feet long," Pavlics said. "The idea was that the moon was covered with a very fine, dusty type of material, and that everything would sink that you put down on it. So, to simulate that, we used flour—and that's a lot of flour in a bin like that.

"We had a lot of problems with rats and mouse," he added.

Another approach was a model fitted with spaced-link tracks, a form of caterpillar tread that Bekker had patented back in Canada. It was an intriguing design, aimed at addressing an age-old problem faced by tanks, tractors, and bulldozers. Traditional tracks are made of hinged plates linked into a continuous belt, onto which are added grousers, the raised cleats that give the tracks their "bite" in soil, sand, and mud. But the plates themselves prevent the grousers from pushing against the soil effectively; they get in the way, so that a track is constantly working against itself. Bekker's solution was to get rid of the plates, with a track that consisted only of linked grousers. In its rover model, the GM team looped spaced-link tracks around the entire vehicle. It churned its way over sand, flour, or gravel.

It was a third, somewhat more conventional model that most excited the Santa Barbara team. Since 1950, when he was working on the Cobra for the Canadian army, Bekker had been intrigued by the "train concept," as he called it: designing vehicles broken into two, three, or more sections, linked by articulated joints. The most familiar style of articulated vehicle to Americans of a certain age might be the accordion-waisted buses that served many big cities in the sixties and seventies; their hinged middles enabled them to negotiate tight intersections that their length would have otherwise nixed. Those buses articulated on just one plane, however—the joints permitted their front and back halves to swing left or right, or to yaw. Bekker recognized that in off-road applications, those joints would have to also permit up-and-down movement, or pitch—and, ideally, allow each half to roll independently from side to side, as well. That way, each piece of the vehicle would be able to respond independently to the ground beneath its wheels, and keep those wheels firmly planted, no matter how rugged the terrain.

A trio of GM's early six-wheeled rover models, parked outside the Santa Barbara lab with a mock-up of NASA's Surveyor lander. Their all-wheel drives and flexible frames lent the models an astounding agility. (ROMANO FAMILY)

He'd taken a step toward full articulation by designing a two-hinge joint, but the demands placed on Earth-based vehicles precluded advancing the idea much further. For one thing, joints that emphasized freedom of movement tended to surrender strength, so in seeking mobility, one traded away reliability and endurance. And to truly take advantage of articulation, all of a vehicle's wheels had to be powered. That wasn't always easy to pull off.

Pondering the matter in Santa Barbara, now with lunar applications in mind, Bekker hit on a new idea: instead of having rigid chassis components connected by overly complex joints, why not make the whole frame flexible? The GM team experimented with a train of three separate modules, each with two independently powered wheels, linked by thin, flexible rods. The result was a model of astounding agility. "It could climb over an obstacle nearly twice as high as the diameter of the wheels," Pavlics said. Because the elastic links between the three pods were free to twist into any combination of pitch, yaw, and roll, the six wheels stayed firmly planted. And because the model's weight was divided among the three pods, the ground pressure created by the wheels was far slighter than that of a conventional vehicle.

Perhaps the setup's greatest advantage was that the three pairs of wheels helped one another. When the model encountered a tall step, its front wheels clawed their way up its face, the rear four wheels pushing them along, and the frame bending to keep all in traction. The front wheels reached the top as the second set of wheels started its climb; now the front wheels, having surmounted the step, pulled as the rear wheels pushed. By the time the rear wheels reached the obstacle, the first two sets were up on top, pulling. Even a partially flexible vehicle worked pretty well, Bekker found. With six-wheel, all-wheel drive, a link between two of the axles could be rigid, and the vehicle would still cross rough terrain with far more aplomb than a conventional off-roader. It wouldn't match the agility of the all-flex version, but it remained impressive.

Another vital advantage to the elastic frame: the rover could be bent like a pretzel to fit into the available cargo hold on a spaceship. "Not until the advent of lunar surface locomotion was the full potential of articulated vehicles, both tracked and wheeled, fully understood by me," Bekker wrote later. "Fortunately, the design of lunar roving vehicles has not been restricted by requirements which must be satisfied on our planet.

"Analysis has shown that an articulated, six-wheeled vehicle is, perhaps, the best answer to the mobility problem."

He and Pavlics bored into the concept, debating both the style of motors best suited for such a vehicle and the design of the wheels. In the years to come, their six-wheeled model would inform many of the lab's concepts for earthly combat and supply vehicles. One took the form of a three-unit train, later lengthened to five, and served as a test bed for articulating joints, power trains, and steering techniques. Another was an articulated, three-quarter-ton amphibious workhorse called the Sidewinder that could carry a cannon, troops, or cargo pretty much anywhere. Santa Barbara also produced an articulated four-by-four that looked something like a pickup with a flexible waist, and which could carry 2,500 pounds of whatever, wherever.

But those were happy by-products. What most interested Bekker and Pavlics was way, way, *way* off-road.

13

PUTTING THOMAS GOLD'S ADVERTISED THEORIES TO THE TEST proved vexing, because America's early attempts to get a closer look at the moon were spectacular failures. Over the course of twenty-eight months, starting in August 1958, the United States launched eight rockets aimed at lunar flybys or orbits. The first, an air force probe, exploded nine miles above Florida. The second, an air force–NASA craft called *Pioneer I,* burned up in the earth's atmosphere two months later. *Pioneers II* and *III* did the same. In March 1959 *Pioneer IV* escaped the earth's gravity—making it the first U.S. spacecraft to do so—but flew too wide of the moon to take pictures. Another flight the following November ended with a bang forty-five seconds after liftoff.

Von Braun's rockets were among those that blew up. He complained that the American style of planning and budgeting was partly to blame. The Soviets set goals for themselves using five- or ten-year horizons, while the federal government's fiscal-year timelines bred crash programs. "I'm in favor of a little less crash," he said, "and a little more program."

Even as he made this comment, he and his Huntsville team were developing a much mightier and more complex rocket, with a far more ambitious purpose. Originally called the Juno V, it would lift off the pad with twenty times the power of a Redstone missile. The Army Ballistic Missile Agency was developing the rocket in-house, following the pattern von Braun and his Germans had established at Peenemunde. They designed, tested, and fabricated the first Juno Vs with their own hands, in their own labs and shops, drawing on the experience they'd gained together on simpler projects. Much about the rocket, big though it was, was familiar to them: the booster stage was essentially nine fuel tanks strapped together, eight based on the Redstone, bundled in a circle around another based on a Jupiter, which was itself descended from the Redstone. This quick-and-dirty design enabled the team to use its existing tooling machinery and workshops, rather than start from scratch. The complication was that the Juno V relied on eight engines working in perfect synchrony, but that is what was required to carry larger communications and military satellites into space.

The team proposed another, secret role for the booster: mated to two upper stages fashioned from missiles, it could deliver a spacecraft to the moon, or serve as a space truck to launch pieces of an even mightier rocket into Earth orbit rendezvous. In June 1959 the ABMA produced a classified report detailing how the army might land men on the moon by 1965 and create a lunar colony by the following year, using sixty-one of the new launch vehicles and eighty-eight of an even more powerful variant.

Project Horizon, this scheme was called. Von Braun had little direct role in it, but several of his Germans were involved, and its four-volume report—which relied on both in-house expertise and input from army technical commands—was typically thorough, figuring the logistical needs for the venture almost down to the pound. It was also preposterously ambitious: with or without von Braun, the U.S. Army was in no way ready to take on such a program, seeing as how it relied on 149 successful rocket launches in little more than two years (at a rate of 5.3 per month), a lengthy lunar construction effort (including a lot of excavation, so that the base could be buried under three feet of moondust), and keeping twelve colonists healthy and supplied for stays of up to a year. Just making the space suits for such an endeavor was probably beyond reach.

For our purposes, Project Horizon was interesting because it envisioned its moon men using two styles of wheeled lunar vehicle. The first was a 4,500-pound machine built for construction; in sketches accompanying the report, it resembled a tow truck crossed with a road grader and married to a bulldozer, its cockpit enclosed in a glass bubble up top. The other was more a pickup, a "low silhouette, skeletionized [*sic*] vehicle of light weight metal construction," with a pressurized cabin for two and a cargo deck behind. Electric motors would turn its single axle, while it steered with a third "tiller" wheel, and it could be linked to other powered axle assemblies as needed—a variation on Bekker's train concept, though minus his flexible frame.

Nothing came of Project Horizon's moon colony, and its vehicle concepts died on the page. Not so the Juno V, however. It went by a different name in the ABMA report, one that the von Braun team had taken to using early in its work on the rocket: the Saturn I. If nothing else, the report advertised its possibilities, and while the army had limited use for such a booster—in fact, U.S. Defense Department policy barred it from using so big a rocket—that wasn't true of the still-aborning NASA. On July 1, 1960, after a lot of slapping and shoving between NASA, the air force, and the army over the Saturn and its German engineers, much of the ABMA was transferred to the space agency, and Wernher von Braun and his team found themselves finally devoting most of their time to civilian spaceships. A giant chunk of Redstone Arsenal's interior, more than 1,800 acres, was carved out around them for the new George C. Marshall Space Flight Center, named for the former army chief of staff, secretary of state, and Nobel Peace Prize recipient. Its mission: to develop the new super rockets, beginning with the Saturn I, that would answer America's growing ambitions in space. Its director: von Braun.

The transfer worked no magic in the ongoing effort to get a ship near the moon. Another probe burned to a cinder in September 1960, and that December, the last of the early Pioneer missions blew to bits at T plus seventy-four seconds. Meanwhile, the first seven American astronauts were undergoing selection and training. One can only imagine how they viewed this parade of woe—and one can understand that when, six months after the last of these failed missions, Alan B. Shepard folded himself into a tiny Mercury capsule for the country's first manned spaceflight, it was attached to a version of von Braun's tried-and-true Redstone missile.

America was again playing catch-up with that modest, suborbital accomplishment: three weeks before, Soviet cosmonaut Yuri Gagarin not only launched into space but completed a full orbit of the earth; by contrast, Shepard's ride lasted just fifteen minutes and twenty-eight seconds from launch to splashdown. Between the two flights, Vice President Lyndon B. Johnson tapped von Braun for his opinions as to how the lagging American space effort might leapfrog the Russians. He answered that it had a poor shot at beating them to an orbiting space station. It had only "a sporting chance" of soft-landing an unmanned spacecraft on the moon first or sending a manned rocket around the moon before they did. But it had "an excellent chance of beating the Soviets to the first landing of a crew on the moon," he offered. "With an all-out crash program, I think we could accomplish this objective in 1967/68."

NASA's top leadership was already leaning toward a moon shot, so whether von Braun's odds-making held much sway is debatable. But speaking to a joint session of Congress a few weeks later, President John F. Kennedy turned that "excellent chance" into a national challenge: "to achieving the goal, before this decade is out, of landing a man on the moon and returning him safely to the earth."

14

THE MARSHALL SPACE FLIGHT CENTER'S ORGANIZATION RELIED ON the same principles von Braun had developed at Peenemunde and installed at the ABMA. At its heart was the philosophy that in a highly technical organization, top-down leadership did not yield good results. "A team is made up of many individuals," he explained in a 1962 speech. "The more individualistic, the better. The smarter the people at the working level are, the better the team. I think nothing hurts a team effort more—and the exploration of space is the greatest of team efforts—than what you might call the 'pappy knows best' attitude on the part of top management. Pappy just doesn't always know what is best. He gets the best answers if he asks the man who is to do the job."

To that end, the Marshall Center was arranged into nine laboratories, each with deep expertise in a facet of space engineering: astrionics (or spacecraft electronics, such as navigation), aero-astrodynamics, computation, propulsion and vehicles, and so forth. Each had "full cognizance and responsibility for all effort" within its specialty, von Braun said. "Competence in depth in each discipline exists at no other point in the organization." Each of the nine was divided into divisions focusing on subspecialties. "The labs were led by Germans, and most of the division chiefs were German, too," said William W. Vaughan, who was one of the first Americans to lead a Marshall Center division. "They knew what they were doing, and we had to learn, to be candid with you."

The fact that the lab directors had worked with one another, and with von Braun, for most of their adult lives lent an intuitive informality to the center's operation. If, in the course of its work, one lab ran into a question best answered by the experts in another, "We simply walked down the hall," Vaughan said, and the other lab was bound to comply. "And if we had an idea, we could always run it past them, or even get them to run it in some test cases to see how it came out."

Designing and building a rocket, or even a small component of a rocket, drew on the strengths of many labs. So overlaid on top of the lab structure were project offices that managed the interdisciplinary nature of the center's complex jobs. These offices often dwarfed the labs, with the Saturn Program Office the biggest of all. "The task of the project office is not to do any part of the technical job in the various disciplines," von Braun explained, "but rather to assure that all effort required by the project has been planned for, budgeted for, and is actually being accomplished in a coordinated, effective, and efficient manner."

To keep all the players tuned in to one another's work, von Braun presided each week at a "board meeting" of lab and project bosses, along with specialists invited for cameo input. Though each meeting followed a formal agenda, it usually gave way to long and often impassioned debate, which von Braun encouraged. When everyone had said his piece (and it was always men doing the talking) the director stepped in to summarize the topic, integrating the various viewpoints, distilled to their essence, into a narrative that everyone present could agree to. The meetings ended in consensus.

To hear his colleagues describe it, von Braun's performance at these meetings bordered on the magical, combining an instantaneous grasp of the most arcane and difficult engineering problems, a gift for translating technobabble into plain language, and a firm but gentle hand in paring away the errant slivers of an argument so that the rest fit into whatever mosaic he was constructing. "The ideal scientist-manager . . . needs a broad background and experience in mathematics, chemistry, physics, and engineering," he explained in the same 1962 speech. "He must understand the relationships among the various disciplines, and their interface in applications to his project. Each specialist sitting around the table must feel that the director understands the problems in his particular field. The director must be able to discuss the problem in the language of that discipline."

Von Braun relied on a low-tech but effective tool to further foster internal communication and team spirit. Late each week, the various divisions in each lab prepared summaries of their activities, as well as any problems they were struggling to solve, and submitted them to their lab director. Each lab compressed these insights into a one-page brief. "Every Friday afternoon, Bonnie [Holmes], von Braun's secretary, collected the weekly notes from those laboratories," Bill Vaughan said. Von Braun "took all those weekly notes home over the weekend, and he read them, and he'd annotate them. He'd write, 'Hans, have Fritz look at this,' or 'Let's have a meeting on this.' Come Monday morning, he gave all those notes to Bonnie. She copied complete sets of the weekly notes, annotated and everything, and gave it to each of the laboratories" and project offices.

Marshall's entire staff sought out copies. Through them, "everybody knew what everybody else was doing," Vaughan said. "You saw the notes from everybody else with his comments on them. There was no competition. There was no secretive stuff." Instead, workers throughout the organization gained a more complete sense of the center's operations and their place in them. "It was a team environment that von Braun created," Vaughan told me. "Everybody was on the team."

15

JOINING NASA REQUIRED VON BRAUN TO ACCEPT ONE PRETTY major paradigm shift. The ABMA style of in-house development—of "keeping our hands dirty," as he described it—would give way, by necessity and NASA policy, to an increasing dependence on private contractors. A good example was the Saturn I rocket: after building the first examples itself, the Marshall Center turned its production over to the Chrysler Corporation. The same would go for the more powerful rockets to come. The center would itself tackle the Saturn V's general design, but once it entered production, the Boeing Company would build the big S-IC booster; North American Aviation, the S-II; the Douglas Aircraft Company, the third stage; and IBM, the instrument unit that married them all into a smoothly functioning whole. Scores of other companies, employing thousands of people, would build the components that fit into these constituent pieces.

We tend, all these decades later, to look back on the Apollo program as NASA's doing, and certainly the agency deserves much of the credit for Apollo's successes. But not all of it. The space program was a public-private partnership well before that term became a business school cliché. Al Haraway, a Boeing engineer in Huntsville who was destined to work on the rover, offered me a nice summation of how the relationship typically worked. "NASA did a conceptual design," he said. "That led to a detailed set of requirements. Those requirements led to a procurement from contractors for elements of hardware. That thing that the contractor proposed was its best analysis of how to meet the requirements that the government had made.

"You then enter a development phase, where [the contractor] says, 'Okay, this is how we're going to get specific about how to go about building these components to meet the requirements. All through, the government maintains a veto power over how the contractor purports to meet the requirements."

Marshall didn't give up its hands-on technical work altogether. Von Braun insisted that its labs stay at least as smart as its contractors. "In order for us to use the very best judgment possible in spending the taxpayer's money intelligently, we just have to do a certain amount of this research and development work ourselves," he said. "We just have to keep our own hands dirty to command the professional respect of the contractor personnel engaged with actual design, shop, and testing work. Otherwise our own ability to establish standards and to evaluate the proposals—and later the performance—of contractors would not be up to par."

His people didn't hesitate to look over a contractor's shoulder. Or to demand that said contractor produce documentation for every single piece it fabricated, every test it ran, every difficulty it encountered. Or to insist that it keep its place clean and well ordered. Or to ride herd on its spending and the pace of its work. And if all that wasn't done to their liking, they were always ready to take over the job themselves, if need be—and that was no idle threat.

16

AMONG THE CENTER'S YOUNG PROJECT MANAGERS—AND ALMOST all were shockingly young—was Sonny Morea. His backstory was not unlike that of many who turned up in Huntsville in its early days as "Rocket City, U.S.A." He was born in Queens in January 1932 to immigrants from southern Italy. As a kid, he was crazy about airplanes—a student of flight at fourteen, soloing two years later, a licensed pilot a year after that. Engineering called. He enrolled in the City College of New York.

As a campus dance approached, he suggested that his girlfriend invite someone to pair with his buddy Bob. She brought Angela Fiore. It turned out that Bob didn't dance, so Sonny took Angela onto the floor. "Tennessee Waltz" was playing. They married in 1955.

After graduation, his dad gave him a used Mercury, and the Moreas drove to California, where he spent a year with North American Aviation before reporting to the army for two years of service. His first orders came as a surprise. "All my buddies went to Korea," he recalled. "When they were passing out assignments, the officer in charge would say, 'Lambert—to Korea.' Guy goes up onstage, shakes hands, takes his orders. 'Johnson—to Korea.' Same thing. It went on and on. Then he got to me. 'Morea—to Korea.' Then he said, 'No, wait a minute. Morea—to Huntsville, Alabama?'

"He didn't know where it was, and neither did I," Morea said. "How's that for a fortuitous accident?"

The couple and their infant daughter drove into town on June 10, 1955. Huntsville was just starting a boom touched off by the burgeoning missile works at Redstone; over the coming decade, the influx of rocket builders and aerospace contractors would transform the city into a cultural center, with an overstuffed arts calendar, its own symphony orchestra, new civic buildings, and a well-respected technical university. At the time, however, the town remained much as it had been when the Germans arrived—sleepy, hidebound, racially intolerant. "It was a shock," Morea said. "I was a kid from New York, and this was Alabama in 1955."

Morea reported for duty to the ABMA's commanding officer. "He told me, 'I see you're a mechanical engineer. We have a group of Germans here that are doing rocket work for us, and they need mechanical engineers. So why don't you go talk to von Braun on Monday morning, and see if he can use you. Then come back and tell me what you want to do.'

"So, I went on Monday morning to see von Braun," Morea said. "He said to me, 'I've got nine laboratories here. They're all headed by a German. They don't speak English well, but they'll understand you, and you'll understand them. I want you to talk to all of them. Find out what those laboratories are doing, whether they require a mechanical engineer, and where you might be interested in working. Come back and tell me, and I'll assign you to do that for the next two years.'"

Mind you, von Braun was a celebrity or close to it, and Morea was a freshly minted army second lieutenant. Few places, it seems, held a bachelor's degree in mechanical engineering in such high regard.

After completing his survey of the labs, Morea went to work first in a division handling thermodynamics. "I wasn't there but for two or three months when one of von Braun's deputies came by my desk. He could look over the barrier—it was a bullpen type of thing—and he said, 'Mr. Morea, you're from New York, aren't you?'" When Morea replied that he was, the German told him that he was having trouble with a Long Island contractor supplying components for the Redstone missile and needed a representative on-site, and a New Yorker would make a particularly nice fit. "I said, 'Sure,'" Morea said. "My wife almost ate me alive because she was stuck taking care of the baby while I was up visiting relatives in New York."

The assignment moved him to ABMA's Guidance and Control Lab. His success at it put him on von Braun's radar. "Pretty soon I got an interest in what was going on in one of the other labs, in propulsion," Morea said. "I was getting out of the service at that time, and it was a convenient time for me to pick another area to work in. Propulsion was headed by another German, Konrad Dannenberg. I went over to see if he had anything he could offer me, and he said, 'Yes, yes. I'd like you to join my office.' They were in charge of developing new engines and modifying existing engines, that sort of thing. That sounded more up my alley."

Before long, the von Braun team's transfer to NASA was at hand. The Marshall Center's new manager of engine programs chose Morea to work first on the H1 engine, which powered the Saturn I and its brawnier offspring, the Saturn IB, and later to ride herd on the fearsome F1 engine, which was being designed and built for the future Saturn V by Rocketdyne, a contractor in Canoga Park, California. The assignment was bounced upstairs to von Braun for approval. "He at least knew who I was," Morea said. "He had no problem agreeing with it. He thought that was fine."

So began Morea's lengthy career as a project manager, which would lay the groundwork for his central role in our story, though that role was still years in the future. "I didn't have to use a slide rule to figure anything out," he said. "I knew the principles of how rocket engines worked and what they needed to do, but I wasn't a mechanic. I was strictly an overall top leader of the program, its executive. My job was to keep track of how the project was going and whether it was being managed properly by the contractor. I had a chief engineer, and a team of people who kept up with what was being charged to the government versus the work that Rocketdyne was performing, and who did the analysis on whether the work was on schedule, and whether there were technical problems.

"So, I never got hands-on in figuring out or fixing a problem. I made sure the right people were working on it, including the right people within the Marshall Space Flight Center, because we had an engineering staff at the center that I could call upon to test hardware for us in the backyard."

One afternoon, Morea took me on a tour of that "backyard." We drove to the Redstone Arsenal, just south of the U.S. Space and Rocket Center, and, after passing through security, rolled slowly through four miles of heat-wilted army post to the Marshall headquarters campus. The eleven-story Building 4200, where von Braun had presided from nine floors up, dominated a cluster of modernist office blocks of steel and glass. The burgeoning Saturn program fast outgrew the old ABMA complex before moving here in 1963, and the new offices fit the mission at hand: they were serious, streamlined, smart.

The center's shops and laboratories occupied the square mile just to the south—low, unadorned shelters, most of them, evidence that NASA was too busy with what went on inside to worry much about curb appeal. Farther south, labs and parking lots gave way to slash pine and bare red earth. Tucked in the trees stood the Redstone test stand, small and spindly, where the missile's engines were fired and tweaked in the 1950s. Then, as Morea slowed his Honda SUV to a crawl, three massive structures came into view.

The first was the static test tower, built in the mid-1950s to test-fire the Jupiter missile's engines and modified later to accommodate the entire first stage of the Saturn I. It amounted to a giant steel and concrete vise to hold the booster fast while its eight engines erupted. Below it, a colossal steel scoop and deep concrete trough channeled the resulting torrent of fire and fury away from the stand and, God willing, anyone or anything nearby. Never a handsome structure, the old tower's steel was now brindled with rust, and saplings sprouted from its upper decks.

Morea next eased the SUV to the Saturn V dynamic test stand. It was in this corrugated metal box that the rocket lying in state at the Davidson Center was shaken for months in the mid-1960s, to ensure that the Saturn V wouldn't lose its bolts or sense of direction amid the stresses of launch and flight. At 371 feet, it was the tallest building in Alabama when it was finished in 1965. It's still the tallest in Huntsville.

Finally, we pulled up to the soot-blackened S-IC test stand. Built to epic scale on four tapering concrete legs that are fused to bedrock forty feet down, this might be the sturdiest piece of architecture in the American South, and not without reason. The Saturn V's booster was strapped into this mammoth harness, and all five of its mighty F1 engines let loose at once. Their 7.5 million pounds of combined thrust would have pulled a lesser building straight out of the ground. Some twenty-two booster tests took place here. They could be heard twenty miles away and sent shivers through the ground under Huntsville. Windows broke all over town.

We parked beside the S-IC test stand for several minutes, eyeing its legs, sturdy as a castle's keep, its soaring steel-girder ramparts. The investment in this largely forgotten monolith was mind-boggling. But this was part of the von Braun team's drive to sort out every last kink in its rockets, to test them until they broke, to fix where their thresholds lay. To keep its hands dirty.

And no piece of the Saturn V was more a linchpin than the F1. If it didn't work, the rocket wouldn't get off the ground, let alone to the moon. Morea was out here fifty-odd years ago, watching those tests. The pressure on him must have been suffocating. "It was a nearly billion-dollar program, and I mean a billion 1960 dollars," he said. "I was twenty-eight years old when I got that job. Can you imagine that?"

17

STILL LACKING HARD DATA ABOUT THE LUNAR SURFACE, THE JET Propulsion Laboratory got to work on reconnaissance programs to follow up the disastrous early Pioneer flights. Owned by NASA but managed by the California Institute of Technology, the Pasadena lab was the country's lead organization for the robotic study of the moon and the planets, and the strategy it devised was supposed to peel back the moon's mysteries one veil at a time.

Its Prospector program aimed to satisfy two important NASA goals: establish the conditions on the moon's surface for a future manned expedition while putting the Saturn I through its paces. It called for planting an instrument-packed lander on the regolith. In 1961, there was talk that it might also carry a remote-controlled rover.

Coming along right behind Prospector was the Surveyor program, with which NASA hoped to check out specific Apollo landing sites. The Surveyor landers were far smaller and lighter than Prospector, but the Jet Propulsion Lab planned to load them with features, such as a television camera that could fire off thousands of pictures of the surrounding environment; instruments to sample and analyze the lunar soil; and, just maybe, a small robotic vehicle.

Aerospace companies didn't wait for either program to jell before rushing to design and build their own moon cars, with future government contracts in mind. The results could be seen, for better or worse, at the American Rocket Society's "Space Flight Report to the Nation" conference, held at Manhattan's New York Coliseum in October 1961. This "astonishing exhibition of the phony and the competent, the trivial and the magnificent," as *Time* magazine described it, included entries seriously vying for NASA's consideration, a great many flights of fancy, and a few that succeeded only in earning scorn. As a visiting engineer commented to the magazine about one display, "It's wonderful what a kid can do with an Erector Set."

The serious efforts involved big money. RCA's Astro-Electronics Division, which typically focused on satellites, experimented with several rover variations, all remote controlled and all of which walked. Some were simple lozenges that waddled on stiff, unsegmented legs. Others overtly mimicked insects, with antennas rising from their "heads" and soil-sampling drills for proboscises. One, with sixteen legs, rowed across the ground in what seemed a cross between a centipede and a Roman galley.

Another outfit's rover stretched out a pair of mantis claws to stab the ground and pull its carcass along. Space-General, a California company, showed off a triangular rover that high-stepped along on six jointed legs, navigating via a TV camera on a long neck while picking up samples with crablike pincers. Legs were the rage at the Coliseum; whatever the challenges of the lunar environment, the bulk of the country's aerospace engineers evidently believed that legs would best surmount them.

Greg Bekker had little patience for the "science-fiction-like models of lunar surface vehicles that walk, jump, and crawl," as he wrote later; the complexity of such devices was "so great, in comparison with a track and wheel," that they deserved "no serious thought." Granted, on a few specific surfaces, a jumping machine might outrun any other, but it would need "an automatic sensing and stability control device comparable in efficiency to the brain and nervous system of a mountain goat."

His own research indicated that the best answers for lunar mobility waited in "rather conventional vehicles." As for the uncertainties of the moon's crust, they weren't that important, given the rather limited range of possibilities. "Although detailed information pertaining to lunar soils is not available, the following assumptions are undoubtedly true," he wrote in 1962. "If there are granular masses on the moon's surface, then the lack of atmosphere and water eliminates soils of the 'plastic' saturated clay type. Only a dry, granular soil of a gravel, sand, or powder may be contemplated." That simplified the designer's job. Even if the regolith turned out to be more like flour than sand, dry soils would "present a rather easy engineering problem, much easier than those faced on Earth in loose mud, where giant tires and large tracks are required."

A rigid wheel would fare poorly, but Bekker reckoned a flexible wheel, mimicking the behavior of a low-pressure pneumatic tire, should do fine. In fact, a flexible wheel forty inches or more in diameter matched the performance of a caterpillar track of the same length. Any smaller than forty, and the track won—but probably not by enough to justify its greater weight and complexity. The choice was actually pretty easy, Bekker decided. "The selection of an economical, less complex mechanism, i.e., the wheel, is the only logical decision under the assumed conditions."

Prospector didn't last long. The weight and complexity of its landers ballooned until they could no longer reach the moon aboard the Saturn I. They would instead require the larger, far more expensive, and breathtakingly powerful Saturn V, which von Braun and his team were now designing at Marshall. Besides that, the program's ambitions grew along with its payloads, until they wouldn't so much lay groundwork for future manned flights as compete with them. NASA killed it in 1962.

It's worth a brief postmortem only because Prospector ushered in a company destined to become an important player in our story. The directors of the multitentacled Bendix Corporation, known for computers, automotive brakes, and a wide range of electronic products, had decided in the fall of 1960 to make "extraterrestrial surface exploration a major, long-term company interest," as one of its reports to NASA said later. Early in 1961, Bendix put together a team "to examine the problems associated with lunar roving vehicles, considering all phases of lunar operations."

Its Aerospace Systems Division, based in Ann Arbor, Michigan, had dreamed up a rover to ride aboard Prospector. Bendix foresaw it spending three months on a slow and steady rumble across the lurrain, racking up five hundred miles and beaming home television images of whatever it encountered. It was a gangly, three-wheeled thing, twenty-two feet long and eight feet high. Its wheels, made of flat aluminum spokes joined by treads of stainless steel webbing, were five feet in diameter and a foot wide. A small nuclear reactor provided its power. It weighed a hulking 1,700 pounds.

Bendix never got around to building it. The closest the company came was a one-tenth-scale, unpowered model. Nevertheless, its concept went public with a bang, landing on the cover of *Aviation Week & Space Technology*, a must-read for those working in aerospace. A photo spread inside showed off the model unfolding from its lander and venturing off into the lunar unknown. The accompanying article hinted that this was no one-off but the first shot in a long and dedicated campaign. So it was: the following decade would see Bendix spend $8 million from its own bank account developing rovers for the moon—when $8 million was serious money—along with another $3.27 million in contracts from the government. No other company put as much cash on the table or so hungered for a capstone contract to justify its investment. And throughout, it would tangle with one rival time and again.

That rival had a new name. In January 1962 the Santa Barbara lab was caught up in a General Motors reorganization, the first of several it would endure during the 1960s and 1970s. It lost its division status but gained some autonomy: now it was known as the GM Defense Research Laboratories, and for a while, Romano, Bekker, and Pavlics had the luxury of pursuing their interest in the moon without much fuss from Detroit or anywhere else.

"We were in an environment where we didn't have a contract, early on, so there wasn't the pressure," said the lab's Paul Jaquish. "We could get halfway through a design, and somebody could say, 'What if we do something like this?' and we could scrap what we were doing to start something new.

"Having come out of the auto industry, meeting production schedules, then going to Santa Barbara, I thought I had gone from hell to heaven."

While Bendix concentrated on its Prospector concept, the GM lab was experimenting with a pumped-up version of its flexible six-wheeled model, a test bed for refinements of the idea. Bekker and Pavlics described it in a May 1963 paper prepared for the company's own staff: It was five feet wide and twelve feet long, with four and a half feet between its axles. The frame linking the axles consisted of six circular rods, each a quarter inch in diameter, arranged side by side, which gave the vehicle a limber surefootedness; rods of greater diameters could be installed if a mission required less elasticity.

Two sets of silver-cadmium batteries powered the test bed. They were split between the front and rear axles, to distribute their weight, and fueled a DC electric motor in each wheel hub. The motors, rated at one-fifteenth of a horsepower each, could manage a top speed of only about two miles an hour, but the torque they produced would push and pull the prototype over practically any terrain.

The GM team tested the machine on the dunes at Pismo Beach, sixty miles up the coast from Santa Barbara, and in the volcanic Amboy Crater near Needles, California, both of which it figured might be earthly analogs, or stand-ins, for the lunar surface. The prototype encountered dunes of loose sand at Pismo, some as steep as thirty-three degrees, that it "negotiated with little or no difficulty," Bekker and Pavlics wrote, while at Amboy it breezed over steep jumbles of gravel and boulder.

Contributing to this nimbleness was the prototype's ethereal weight. "In order to simulate a 400-pound (Earth weight) vehicle performing under conditions of lunar gravity, the gross weight of the test bed has been held to 66.6 pounds," the authors wrote. The vehicle had no body to speak of (this was a test bed, after all), and its wheels, three feet in diameter and fifteen inches wide, were made of closely spaced wire hoops connected at both ends to the rim, then covered with a polyester cloth. The hoop construction recalled an 1858 English patent, also reliant on wire; it enabled the GM wheels to deflect and absorb impacts much like balloon tires. Their own feathery weight, combined with the light load they carried, graced the prototype with a ground pressure of only 2.4 ounces per square inch. It practically floated.

The pair would not return to the nineteenth-century style of wheel, per se. But in giving the old idea a try, they discovered the

Frank Pavlics, standing at left, and Greg Bekker pose with a pair of their early rover models alongside a mock-up of a Surveyor lander. The machine at left is sized for delivery aboard Surveyor; the other resembles the larger test bed that demonstrated the advantages of their six-by-six, flexible-frame concept. (FACULTY OF AUTOMOTIVE AND CONSTRUCTION MACHINERY ENGINEERING, WARSAW UNIVERSITY OF TECHNOLOGY)

merits of wire, and that they did use again, many times. It was light. It was springy. And, depending on what kind of metal it was made of, it could be tough enough to stand up to the moon's hard vacuum and five-hundred-degree temperature swings.

"The lunar roving vehicle concept has displayed unusual mobility in negotiating both soft, loose surfaces and rough, undulating ground," Bekker and Pavlics wrote of the test bed. "If the lunar surface is found to be as rough . . . as many investigators think it will be, it is entirely possible that roving vehicles based on this concept will prove to be superior for lunar exploration."

18

IT WAS LIKELY SOMEWHERE NEAR THIS JUNCTURE THAT PAVLICS happened on a new and elegant invention in his regular review of technical literature, a transmission that could "step down" a fast-spinning electric motor so that it produced the power and speeds needed to turn heavy machinery. Or, as was instantly apparent to him, vehicle wheels.

A small electric motor turns quickly but has little muscle; a transmission reduces its speed but at the same time converts its output into useful power. Traditionally, transmissions relied on busy gear boxes to perform this conversion. They were complex, took up space, and weighed a lot—all of which were strikes against them in space applications, where simplicity, small size, and light weight were essential— but the engineering world had offered little in the way of alternatives. Until now. In place of a clockwork of spinning gears, this new device had three parts, just two of which moved. It was small. It was light. Its inventor called it a strain wave gear, or harmonic drive.

That inventor, Pennsylvania-born C. Walton Musser, held patents for a recoilless rifle, an aircraft ejector seat, and an automatic inflator for life vests, along with a lot of other military gear. In 1955 he was working under contract for the United Shoe Machinery Company of Beverly, Massachusetts—which, despite its name, was a major defense contractor—when he hit upon his harmonic drive.

To explain it in the simplest possible terms, the device attaches an elliptical rotor to the end of a motor's output shaft. This rotor, which

Musser called a wave generator, is the first of the drive's three parts, and it spins, as you'd expect, at the speed of the motor. In the case of those envisioned for GM's lunar concepts, this was somewhere in the neighborhood of ten thousand revolutions per minute.

The motor shaft and wave generator slip into the mouth of a flex-spline, which is best imagined as a cup turned on its side. This cup has thin, elastic walls and a solid, stiff bottom. Its mouth is just wide enough to fit around the rotor, but only by stretching to assume its elliptical shape, so that once the rotor is inside, the cup's sides bulge at two points, opposite each other. Once the motor is switched on, the cup's bulges move with it. One more thing about the flexspline: around the exterior circumference of the cup, right where it bulges, is a ring of gear teeth.

Now for the drive's third and final piece: the circular spline, a rigid ring just wide enough to fit around the bulging flexspline. It is toothed on its interior surface. So, we have a flexspline, ringed by teeth facing out, inside a circular spline, with teeth facing in. The fit is snug enough that their teeth mesh, but only at two points: the bulges created by the elliptical rotor. The drive's secret sauce is one seemingly small design detail: the flexspline has a few fewer teeth than the circular spline. For every rotation of the wave generator, the flexspline and the circular spline change their positions relative to each other by that number of teeth.

In other words, for each full revolution of the motor, the harmonic drive moves just a fraction of a revolution, that fraction depending on the number of teeth on each spline. Installed in a moon car, it can transform a motor's 10,000 revolutions per minute into 125 turns of a wheel, turns with power behind them. And it accomplishes this re-duction almost passively. Because it had so few moving parts, Pavlics understood that it promised long life and high reliability. And because it could be sealed into a single unit with an electric motor—the better to protect the entire drivetrain from dust, heat, cold, or the absence of air and moisture—the drive seemed tailored to space. "I think we saw it as obvious," the Santa Barbara team's Paul Jaquish told me. "These were good engineers at GM, and I think we all thought, 'Oh, yeah, this is neat. We can use this.'"

So now began a years-long collaboration between the GM lab and United Shoe Machinery—discussions about possible uses for the drive, followed by experiments with the gizmo. One choice GM faced was the material for the flexspline. The patent suggested it could be "an elastic material . . . such as rubber, synthetic rubber, nylon, or other plastic, or a metal such as steel, bronze, or other gear material." The trick was that it had to be thin and pliant enough to readily deform around the spinning wave generator, but tough enough to do so countless times, at high speed, meshing with the circular spline all the while, without losing its shape or cracking. Pavlics favored steel alloys; they could be machined to various thicknesses within the same small piece of gear, so a single piece could yield the pliability needed for the flexspline's walls and the stiffness required in the cup's bottom.

"It was a unique drive," Pavlics said. "We needed an eighty-to-one reduction, and this did it in one step, instead of with a bunch of gears. It was small and could be hermetically sealed with the motor. And it was very, very light."

19

MEANWHILE, IN JUNE 1962 NASA DECIDED HOW IT WOULD GO about putting a crew on the moon, thus ending a long-running debate within the agency. Three options had been in play. The first was Earth orbit rendezvous, the method that von Braun had described in the *Collier's* series: two or more Saturns would lift a moon ship and its fuel into low Earth orbit, where the craft would be topped off and bid adieu. After landing on the moon, the same ship would carry the astronauts back. Von Braun and his Marshall Center Germans had been staunch in their support of the option.

The second, and simplest, was direct ascent, in which a spacecraft would be fired directly from Earth to the moon, land there, then blast back off for the trip home. Early on, the top brass at what was soon to be the Manned Spacecraft Center, by now under construction in Houston, backed the approach. It suffered from several disadvantages, one of them major: the moon ship would have to be fairly large to haul enough fuel for a round-trip. For the outbound leg of its voyage, half that fuel would be dead weight; on the return, the ship itself would be heftier than necessary. All of that translated into a pile of mass to get off the pad and into space, meaning a bigger launch vehicle. Since 1960, NASA had studied a superbooster for the job, a monster rocket called the Nova that would dwarf the Saturn V in size and power.

The third option seemed a long shot in the early days of the space program. In lunar orbit rendezvous, a single rocket would propel a coupled mother ship and a tiny lander to lunar orbit, where they'd separate; while the mother ship loitered there, the lander would descend to the moon's surface. At the end of its visit, the lander would launch back into orbit and reconnect with the mother ship. The lander would be discarded before the astronauts headed for home. At each step, the spacecraft would shed mass, so that it carried only as much fuel as it needed for the next. The plan's chief advocate was an engineer at NASA's Langley Research Center named John Houbolt; he had bucked the agency's chain of command to wage an insistent crusade on its behalf.

The choice was thorny. Earth orbit rendezvous required multiple launches—a gamble right there—and a moon ship good for both landing and travel across a half million miles of space, which raised serious design issues. Direct ascent appealed in its straightforwardness, but its reliance on an entirely new booster wasted the work completed thus far on Saturn and put a landing by decade's end in serious doubt. Both of those options required setting down a very large spacecraft on the lunar surface—and seeing as how work was already under way on the Apollo command module, the feat would have to be accomplished while the astronauts lay on their backs, facing the heavens rather than the ground. Meanwhile, lunar orbit rendezvous was the most efficient option, but it called on the crews of two spacecraft to find each other and safely dock. Nothing like that had been attempted in Earth orbit, let alone so far from home.

But at an all-day meeting of NASA leadership at the Marshall Center, von Braun shifted his allegiance to Houbolt's model, clearing the way for work to start in earnest on the systems that would get America to the moon. The Marshall chief explained his change of heart in eleven pages of remarks, which we need not dissect here except to note that it was due, in part, to (a) lunar orbit rendezvous' reliance on a single rocket far smaller than the Nova, namely the Saturn V; (b) its solution to the lunar landing problem; and (c) what he foresaw for the lunar program after the initial moon landing—when he hoped to see astronauts supplied by a second Saturn V, serving as a "logistics vehicle," or space truck. Evidence suggests, in fact, that his NASA bosses pretty much promised that he'd get his long-cherished space truck if he came around.

Von Braun had some interesting uses in mind for the truck, including gas station. "It appears that the [lunar module], when refilled with propellants brought down by the logistics vehicle, constitutes an ideal means for lunar surface transportation," he wrote. Using the lander as a "lunar taxi" would give moonwalkers a "radius of action" of at least forty miles: "It may well be that on the rocky and treacherous lunar terrain the [lunar module] will turn out to be a far more attractive type of a taxi than a wheeled or caterpillar vehicle."

Using the lander to hop around the moon's surface never came to be. But von Braun would hold on tight to the idea of sending a logistics vehicle, and it would color the Marshall Center's work on lunar rovers for years.

20

ALL THROUGH 1962 AND INTO 1963, BOTH GM AND BENDIX KEPT AN eye on the Surveyor program. Sure enough, come summer, the Jet Propulsion Laboratory laid out its requirements for a hundred-pound, remote-controlled rover that it wanted to stash aboard the landers. The vehicle would explore the lurrain up to a mile from the Surveyors, while its drivers back on Earth steered it with television eyes. The laboratory alerted companies planning to bid on the phase 1 design study—the normal first stage of any new hardware program—that they'd be expected to supply engineering models of their concepts. Proposals were due in seven weeks.

The short deadline weeded out the dilettantes. In October the two companies left standing—GM and Bendix—started work under contract. GM was ready with its six-wheeled design. Its Surveyor lunar roving vehicle was six feet long on eighteen-inch wheels and weighed ninety pounds—half the size and half again as heavy as its test bed, with a sure-footedness that was no less jaw-dropping. On Pavlics's "lunarium" of rocks, craters, and slopes outside the Santa Barbara lab, it climbed forty-five-degree inclines, leapt twenty-inch crevasses, and bent its way up and over thirty-inch steps.

Bekker and Pavlics had been working on the idea for more than three years by then. Their main advancement this time: the wheels. Again, they were made of wire, but it was knotted into a wide mesh that resembled chain-link, and shaped into fat doughnuts. Like the team's earlier wire tires, they deflected when they hit an obstacle and absorbed some of the bumps of cross-country travel. They worked with or without a fabric covering.

"We had a big program to try to come up with the wire material that would survive the vacuum environment on the moon," John Calandro recalled. "Frank had devised a testing device that created the vacuum environment we needed."

When fully geared up for a mission, the rover would be an electronic wonder, with subsystems supplied by RCA Astro-Electronics and by AC Electronics, a GM division in Milwaukee: it would have a stereo TV imaging rig, sophisticated navigation and control, and silver-zinc batteries recharged by solar panel. But Santa Barbara's part of the job, the vehicle itself, was a study in doing more with less. The hardware was constantly "assessed to see if something simpler might be able to do the same job," designer Norman J. James would remember. "'The part that's left off never breaks' was an often-repeated phrase."

Bendix took a radically different approach. Its SLRV was a squarish, two-part, articulated robot, with curving, shock-absorbing legs at its corners that ended in small caterpillar track assemblies. The tracks pitched independently to follow uneven ground. Its handlers steered it with commands to slow, speed up, or reverse the tracks on one side or the other, and the pivot linking the two halves did the rest. On the moon, it would be powered by a radioisotope thermal generator—a small nuclear device—hanging off the back, and bristle with scientific instruments and antennas. It weighed one hundred pounds.

Side by side with the GM model, the Bendix machine looked bulky and awkward, and those tiny tracks didn't seem much of a match for Pavlics's nearly spherical wire wheels. But Bendix was bullish on its design right up to the day in May 1964 when a panel from the U.S. Geological Survey, Caltech, and NASA took the two models to a volcanic field north of Flagstaff, Arizona, and turned them loose on the rugged Bonito Lava Flow. "We had one little section where they could really get into some pretty rough stuff," the Geological Survey's Jack McCauley recalled years later. "The GM vehicle was perfect. It got from point A to point B without any mishaps or turning over.

"The poor Bendix vehicle had tanklike treads that were made of

Bendix's Surveyor LRV model, shortly before it self-destructed during its showdown with GM's six-wheeler in Arizona. (USGS)

some kind of rubber-type thing," McCauley said. "The vehicle just started shredding the treads. In fact, when they finished halfway down the course, it had no treads left. So, the GM thing obviously got our blessing."

General Motors had scored a decisive victory. Unfortunately, it didn't add up to a rover on the moon. The "Rover Boys," as that panel of testers came to be known, were mightily impressed with the six-wheeler, but its capabilities didn't square with the Jet Propulsion Laboratory's requirements: namely, to "go around and take pictures every ten meters, and also to use a penetrometer to see what the strength of the lunar soil was—and to do it in a preordained manner," McCauley said. "Basically, just do a grid survey." Bendix had produced too little rover for the mission; GM had produced too much. The Rover Boys reluctantly reported that neither rover matched the Surveyor program's stated needs, and that was among the reasons that NASA scrubbed the rover component not long after.

By that time, JPL's Ranger program had finally given NASA its first close looks at the moon. By design, they were fleeting glimpses: Ranger probes crashed into the lunar surface while taking high-resolution photos right up to the moment of impact. Conceived in 1959, the program had, at times, seemed another exercise in frustration. After Rangers 1 and 2 made two development test voyages in 1961, along came Rangers 3 through 6, all of which were busts. It wasn't until July 1964, and Ranger 7, that the program literally hit pay dirt. As the spacecraft fell toward the moon, its cameras kicked on, and, for some seventeen minutes, it took and transmitted photographs of the approaching surface—4,316 images in all, some of them at a resolution hundreds of times greater than the best taken from Earth. The photos didn't put to rest the fears inspired by Thomas Gold's writings and lectures, but they did establish that the maria were smooth enough for a landing.

21

GM AND BENDIX WERE COMPETING ON OTHER LUNAR PROJECTS even before the Arizona test-drive. In 1963, six years before the first moon landing, NASA was already sensitive to the need for something to follow it, an even more ambitious program that would capitalize on the country's great investment in Apollo. Von Braun, in particular, recognized that keeping the Marshall Center's workforce employed required a continuing demand for rockets, and thus a plan for follow-on missions. To that end, the center awarded contracts for a series of studies on longer-term lunar stays.

One concept, the Apollo Logistics Support System, called for creating a pressurized mobile laboratory, or Molab, in which two astronauts could live in shirtsleeves for up to two weeks while embarked on a 250-mile scientific odyssey. The Molab would be flown to the moon on its own Saturn V space truck. As described by Bendix, it would be "disembarked and checked out remotely from earth." Its drivers would later land in a standard lunar module, and the Molab "driven by remote control to the astronauts."

This "dual launch" approach to putting heavy equipment on the moon traced straight back to the *Collier's* stories—and, of course, built on von Braun's remarks supporting lunar orbit rendezvous. In the spring of 1964, the Marshall Center awarded parallel nine-month contracts to Boeing and Bendix to study the Molab idea and drum up some initial designs. Boeing hired GM as its subcontractor on the vehicle's mobility and navigation systems.

At first glance, the resulting two concepts, unveiled in the spring of 1965, seemed to have much in common. Both resembled the illustrations of von Braun's tanklike moon cars in *Collier's*, only with wheels instead of tracks. Their cabins took the form of thick-walled aluminum cylinders laid horizontally. Each had side-by-side seating up front, and an airlock at its rear so the crew could get in and out without depressurizing the entire cabin. Both were powered by hydrogen-oxygen fuel cells. And both were big and heavy, Boeing's especially so: more than twenty feet long and ten feet wide, the six-wheeled beast threatened to weigh four tons when loaded with astronauts and gear. Bendix's was a touch smaller—its cabin, designed by a subcontractor, Lockheed Missiles and Space Company, was seven feet, four inches across. Weighing in at an estimated 6,800 pounds, it was a hoss, just the same. Even on the moon, these things would not tread lightly.

In the details, Boeing took more chances. Its Molab had a four-wheeled, rigid frame beneath the cabin, but a flexible frame to its separate, two-wheeled rear unit, on which rode two big, spherical tanks for the liquid hydrogen and liquid oxygen feeding its fuel cell. That cell powered electric motors in the six wheel hubs; their speed was stepped down with harmonic drives.

The combination of cabin and fuel tanks gave the Boeing Molab the look, as one observer put it, of "a twelve-hour cold capsule pulling two aspirins." Because it boasted six-wheel drive, the rear unit could push the whole Molab along, if need be, and promised to give the vehicle a bizarre agility, despite its bloated mass. On five-foot wheels, Boeing claimed, it could climb seven-foot vertical walls and span eight-foot crevasses.

The Bendix concept was a four-by-four. Lacking the Boeing's third set of wheels and flexible frame, its climbing ability was half that of its competition. And it wasn't as pretty a vehicle, if that word could be used to describe either; it lacked the fit and finish, the smooth edges, of its rival.

But what most conspicuously set the two Molabs apart were the wheels themselves. Rubber tires were never an option. Though weight wasn't the factor it had been in the Surveyor project, the Molab would be exposed to temperature extremes and solar radiation that would likely destroy rubber tires long before they'd completed the vehicle's two-week mission. Air-filled tires of any material were out of the question, because a flat could maroon astronauts miles from their lander. The challenge was to find a wheel that blended the cushion of a pneumatic tire with bulletproof strength and reliability.

Pavlics supplied the Boeing's wheel, which again consisted of a wire tire joined to a lightweight rim—only this time the chicken-wire airiness of the Surveyor tire was pulled into a tight weave. The lab conducted a raft of experiments on the design. "I did work on the wheel, as far as helping to screen various metals for the wires," John Calandro told me. "I still have scars on my hands from trying to build a wheel out of clock-spring steel. We never could get it pliable enough to use. What turned out to work best was music wire."

GM's Molab concept, here sketched by designer Norman James, was tubby enough to require its own Saturn V to reach the moon—but the flexible link between its middle and rear axles made it surprisingly limber. (NORMAN J. JAMES AND GENERAL MOTORS)

Stainless steel piano wire, that is. Also used in springs, which hints at its fitness for duty in a wire-mesh wheel: it was stiff but pliant, so that it bounced back from a blow. For the Molab, intersecting wires were no longer looped around one another—that would stress the metal far too much. Given the vehicle's weight and the unevenness of the ground it would cross, the wires would snap if they weren't permitted small movement. Instead, they were crimped at regular intervals and crossed at those points. The trick, Pavlics explained, was that they had to be crimped carefully, to create an inflection in the wire without crushing it.

Once woven into a mat, the wire was bent into a cylinder, and its edges joined. Heat treated for strength, curled into a torus—a doughnut shape—it was then mounted on a spun-aluminum hub. While its Surveyor rover wheel had been nearly spherical, GM's finished wheel for the Molab looked more like something off an American sedan, except that you could see through it. That, and it stood five feet tall.

The jumbo wheels envisioned by Bendix had no tires. They consisted of a rigid, spoked aluminum hub, around which were clustered springy circles of titanium. Wrapped around these hoop springs was a smooth band of the same metal. When the wheel encountered an obstacle, the outer band and the hoop beneath it deformed, then sprang back to their default shape. On appearance alone, the GM wheel design trounced its rival. The mesh looked *engineered,* space ready, while the hoop-spring wheel seemed a throwback to the nineteenth century—heavy and wagonish, as if fashioned by blacksmiths.

Which is academic, really, because neither company built its Molab; NASA decided to put off deploying a pressurized, rolling moon base until 1975 or later. Marshall engineers continued to explore the idea, all the same. In the spring of 1965, the center's deputy chief for bioengineering, Michael J. Vaccaro, spent time in a Bendix Molab mock-up at Lockheed's complex in Sunnyvale, California, checking out the machine's ergonomics and equipment arrangement. That summer, he and Marshall engineer Haydon Y. Grubbs Jr. tried on a wooden mock-up of the Boeing rig in Seattle. Then, the following February, Vaccaro and Grubbs were enlisted as guinea pigs for a more onerous experiment: spending the equivalent of an entire Molab mission in and around a mock-up dubbed Lunex II, to determine just how well two astronauts might get along over time in a space sixty-five inches high, six feet wide, and seven feet long—or, as press accounts put it, "about the size of three telephone booths." The pair sealed themselves inside without word on when they'd be freed. Each day they followed a busy schedule of scientific, navigation, and simulated driving duties inside the vehicle, venturing outside only in pressurized space suits to briefly tend to geologic sample gathering and equipment maintenance. The study's overseer, Honeywell's Systems and Research Division in Minneapolis, conjured a lunar ambiance as best it could in the basement of its lab.

The pair stayed in Lunex II for eighteen days. The logbook Vaccaro kept hinted at the stresses of the job. They cracked their heads on the low ceiling. When he fell ill on day sixteen, there was little room for him to lie down while Grubbs tried to work. Honeywell threw simulated emergencies at them—such as on day seventeen, when they were alerted to an imagined pressure leak "while I was on the throne seeing to nature's call." They were always hungry; even so, Vaccaro passed on the freeze-dried beef stew. Finally, the Lunex's aesthetics began to grate on him. "Color scheme could be improved," he wrote. "Brown and gray depressing colors."

When they were finally released, bearded and grimy, their wives were waiting—and so was von Braun, who called the event "the can-opening." Both men reported they could have lasted at least another week and that they were too busy to work each other's nerves. Years later, though, Vaccaro admitted: "We were good friends at the beginning of the test, but we weren't so friendly toward the end."

That was about as close to reality as the Molab program got. On its own, General Motors did build a vehicle that is often mislabeled as a Molab prototype. Actually, though, the mobile geological laboratory was an earthbound four-by-four, built under NASA contract for use by the U.S. Geological Survey. The Survey's scientists spent days in the Arizona desert designing procedures and tools for the astronauts to use on the moon, and NASA figured that a go-anywhere, eight-ton, instrument-laden motor home might prove useful, especially for trips on which the geologists strategized how to best use a vehicle to get around the lunar surface. Also called the lunar mission development vehicle, it had an engine from a Chevy Corvair, an articulated chassis, and a swanky, futuristic cabin with lots of glass, and was apparently pretty comfortable.

It cost $600,000 to build in 1965. Adjusted for inflation, that's just shy of $5 million today. After three years, however, it was clear that Molab was dead, and the Geological Survey gave it back. It wound up at the U.S. Space and Rocket Center.

22

ON FEBRUARY 3, 1966, A SOVIET ROCKET DROPPED A PROBE CALLED Luna 9 onto the Ocean of Storms, the largest of the moon's maria. Upon landing, it did not sink. Its transmitted photos, intercepted in the West, depicted the spacecraft sitting on a plain pocked by craters and littered with rubble, including substantial rocks that likewise had not been swallowed by loose dust.

Four months later, on June 2, the Jet Propulsion Lab's Surveyor 1 soft-landed in the same mare and, over the next six weeks, transmitted more than eleven thousand much sharper photographs—including images that showed the lander's own feet hadn't sunk more than an inch or two into the regolith. While Luna 9 had been a tiny craft, the Surveyor weighed 650 pounds. The lunar crust was firm enough to land on. And, it seemed, to drive on.

Back on planet Earth, von Braun hadn't given up on the idea that moon missions might someday include a space truck. Although NASA had deferred any serious consideration of the Molab program, the Marshall Center contracted General Motors and Bendix to build full-scale, working models of their Molab chassis and drivetrains—including motors, transmissions, suspensions, steering, and wheels. These mobility test articles, or MTAs, were supposed to weigh about one-sixth the weight of the complete Molabs, so that their builders and NASA could analyze how the finished products would behave on the moon.

At 1,800 pounds, the GM entry was Santa Barbara's biggest iteration yet of Bekker's six-wheeled design, only this time with an only partially flexible frame and the latest version of the wire-mesh wheel. Like its smaller forebears, GM's mobility test article acquitted itself well in tests. With one hiccup: at the army's Aberdeen Proving Ground in Maryland, it tore through its hard-surface trials, and took on obstacles and crevasses that outmeasured its wheels. But a subsequent series of desert tests at the Yuma Proving Ground in Arizona thrashed Frank Pavlics's tires: the wire mesh snapped in collisions with rocks, and trips through a boulder field left it with permanent dents. As a stopgap to see the machine through the testing, the Santa Barbara lab added rubber inner tubes inside the wheels, but got busy looking for lasting solutions.

The Bendix model, meanwhile, lumbered through the tests intact, though its performance was only so-so: its eighty-inch wheels were good only for climbing a forty-nine-inch step and crossing a sixty-eight-inch crevasse. It topped out at 7.6 miles per hour on concrete. It resembled an outsized buckboard—an impression bolstered by those wheels, which looked straight out of the Old West. And at more than twenty-four feet long and nearly eleven feet high, it dwarfed the GM vehicle. Von Braun had the chance to drive it in Huntsville. He was not at all a small man, but up at the controls, he looked like a toddler in a suit.

Alongside their hefty Molab designs, Boeing and Bendix were contracted to design a second payload for a space truck, consisting of a pressurized shelter for a two-week lunar stay and an open one-man runabout: a Molab that didn't move, in essence, with a go-cart that did. The name the Marshall Center came up with for this smaller vehicle was local scientific survey module, or LSSM, which was unsexy even by the standards of an agency that called its Apollo space suits "extravehicular mobility units."

Von Braun had signaled NASA's interest in a smaller rover in February 1964: "For short-distance travel, a nonpressurized 'moon jeep' may suffice," he told the readers of *Popular Science*. "The astronauts would hop onto its open platform and depend for protection upon their pressurized space suits, while life support and communication would be provided by their back packs." He'd contracted Huntsville's Brown Engineering Co. to build such a vehicle using off-the-shelf parts, on oversized wheels and pneumatic tires, to check out

Frank Pavlics steers GM's six-wheeled, partially flexible MTA through a shakedown trial in Santa Barbara. (ROMANO FAMILY)

The king-size Bendix MTA rumbles over a muddy course at the Marshall Space Flight Center. (NASA)

the practicality of the idea. Brown's tiny test mule affirmed that an open moon jeep could work.

This was a departure from pretty near every depiction of lunar travel, in fiction and fact, in print and on film, over the prior sixty years. Hollywood spacemen *always* rode around in pressurized cars—or, at the least, in vehicles that had sides and a roof. In contrast, the requirements for the LSSM were minimalist, as they had to be if it were to squeeze alongside a shelter on the space truck's lander. NASA expected it to carry a single astronaut, plus seven hundred pounds of scientific payload, up to five miles from base camp.

Once again, GM Defense Research Labs supplied the mobility system for Boeing's entry. Its LSSM was a shrunken version of its mobility test article: a six-wheeled, semiarticulated rover thirteen feet, four inches long, riding on forty-inch wheels. Its load was divided between its two-axle front unit and its single-axle aft unit, which were connected by a flexible link. The front contained the crew station; the rear hauled the power plant, drive system electronics, communications gear, and navigation equipment. Fully loaded, it would weigh about 2,200 pounds.

Bendix chose to go with a short chassis, with the wheels attached to stalks jutting fore and aft. The design pushed its points of contact with the ground to the vehicle's far corners, theoretically making it more stable and maneuverable. It also enabled the LSSM to fold easily. In the finished version, the front stalks would swivel rearward, and the rear forward, to create a smaller package.

Though not, it should be said, all that small. This junior version of the company's mobility test article had a 154-inch wheelbase—an epic size shared by the modern "Grave Digger" monster truck, and dwarfing the grandest Cadillacs to ever roll off an assembly line. Its hoop-spring wheels were 45 inches in diameter and 10 inches wide. Despite those dimensions, its performance was modest: it could climb over a 19-inch obstacle and cross a 41-inch crevasse, at a top speed of just 5.6 miles per hour.

Von Braun and his lieutenants were gung-ho about the LSSM's prospects. But in December 1965, after learning of the high cost of actually building either machine, NASA headquarters ordered the center to scale down its ambitions by deleting some of the roadster's mandated capabilities. In June 1966 Boeing and Bendix each were awarded a six-month contract to explore a stripped-down "Specified LSSM."

Both companies cut the navigation and communications gear from their concepts, and with them, about two hundred pounds. Boeing's didn't change much in appearance. Bendix lengthened the wheelbase of its design by eight inches and downsized the hooped wheels a touch, to forty-three inches in diameter. The company was satisfied with the outcome. "Results of this program indicate that indeed a simplified LSSM system can be designed without compromising the basic mobility performance goals," it declared, "and that a significant reduction in program cost can be achieved."

Bendix was sufficiently pleased, in fact, that it built a full-scale, motorized mock-up of its Specified LSSM, on which it offered rides to the press and visiting NASA officials. "As motorcars go, it's not at all sleek, streamlined, handsome, sporty, or in any other way virile," a reporter for the *Detroit Free Press* wrote of the vehicle in May 1967. "The first thing you think of is a tractor—an awkward, new-fangled model probably concocted by a government committee."

The reporter was especially moved by the Bendix wheels. "The weirdest-looking things on this weird-looking vehicle are those wheels," he decided. But they worked: when he took the controls and drove over scattered lumber, he didn't feel so much as a bump. While under way on rugged ground, each of the wheels "sort of folds itself over a rock," he wrote, "and then springs back into shape as if it were some kind of sponge."

23

WHILE WORK ON THE SPECIFIED LSSMS CHUGGED ALONG, VON Braun's NASA team was taking a closer look at the wheels it hoped to one day put on the moon. It subjected eight contractor designs to a series of on-paper analyses and decided that only two deserved further consideration: those proposed by GM and Bendix. The Marshall Center requested proposals from both companies for a Lunar Wheel and Drive Experimental Test program, making it clear that just one would win the contract. The winner, in other words, would test and evaluate its rival's wheel and drivetrain, as well as its own. What might seem an unfair leg up to the winning bidder wasn't really much of an advantage, because NASA could always choose to have Boeing build its Specified LSSM with the Bendix wheels, or vice versa. It could choose to pair the Bendix chassis with the GM drive. It could, if it so wished, choose to meld the two approaches completely. It owned the work.

GM won the contract. Its marching orders: evaluate the competing concepts, in their LSSM configurations, on their traction and stability in varying soils; assess the strength and durability of their materials; and measure their endurance in lunar conditions. The showdown pitted differing philosophies. GM intended its wheel to perform its way out of trouble with the attributes of a pneumatic tire, while Bendix strove for simplicity.

The exercise came along as Frank Pavlics refined his mesh wheel to address the weaknesses made plain at the Yuma field tests. The first was the wheel's tendency to overdeflect—to cave inward when it hit an obstacle and fail to spring back to its original shape. Among the vexations of designing for the moon is that while everything there weighs just one-sixth of what it does on Earth, the dynamic forces created by collision are every bit as strong as they are at home. Smack into something, and you're going to feel it.

"Frank told me, 'Look, we need something to keep this thing from overdeflecting,'" John Calandro recalled. "So that was my assignment." His solution was a smaller metal tire within the wire-mesh tire: a wide band of titanium supported by a radial arrangement of stiff titanium hoops, which were riveted to the wheel's aluminum rim. This "bump stop" enabled the mesh tire to deform and absorb the forces of collision, but only so much before the more stubborn inner hoops were called into play. The bump stop would deflect under the force of an especially hard blow, but only by an inch or so.

The other key issue concerned the mesh's contact with the ground. Pavlics wanted to improve the wheel's flotation, so that it rode on top of loose soil, rather than tilling it. And as the broken wires at Yuma had advertised, the wheel's contact surface needed extra protection. A tread, of sorts, seemed the solution. As the lab struggled to devise some way to weave a reinforcing tread pattern into the mesh, GM designer Norman James suggested they simply rivet strips of Vespel, a tough epoxy composite, in a chevron pattern to the tire's contact area—an easy fix if they did the riveting while the mesh was laid flat. Pavlics improved on the idea by breaking each of the strips into several segments, attached with their own rivets, so that the strips and wires could pantograph, or shift enough under a load to avoid straining the weave. He recognized the improvements as dramatic; he shared the patent for the wheel with Calandro and James.

"We had a wheel shop with about half a dozen people working, and weaving the wires together," Pavlics told me in his Santa Barbara living room. "First cutting it to length, then making for every intersection the deformities," or crimps. "We had a machine to make those." He paused. "Here—let me show you." It hung on a wall in his office, stainless steel gleaming dully, surrounded by framed and autographed photos of astronauts. This was the finished rover wheel, subtly different from the version that emerged from the Lunar Wheel and Drive Experimental Test program. The wires were the same, however, and it was the wires he wanted to show off. The mesh, hand loomed at the Santa Barbara lab, consisted of eight hundred wires per tire, each 0.84 millimeters thick, or about one-thirtieth of an inch. Any thicker a wire would mean fewer per tire, boosting the loads on each; any thinner would increase the number of wires and intersections, further complicating the wheel's diabolically complex assembly. Each wire had to be mated with another on the mat's far side to form a wide cylinder, which was then curled into a torus and attached to the rim.

I studied the wheel for several minutes, searching for any telltale bulge or other change in the wire that marked the joints. I found none. "You can't see them," Pavlics told me. "It's continuous. It's absolutely not visible, where the wires are joined." He ran his hand over the tire's curved sidewall. "I've looked, many times, and I can't find them."

The GM test team strove for objectivity in assessing the Bendix wheel and its own. To do otherwise would have violated NASA's intentions and backfired on the company. In the end, its research demonstrated that each wheel offered advantages over the other, and that both did well—though GM's own design held the edge. In one endurance trial, each was mounted inside a spinning squirrel cage lined with diamond-plate steel, into which obstacles of different shapes and sizes were randomly deployed. The device fit inside a thermal-vacuum chamber, enabling the testers to heat the wheels in a hard vacuum, mimicking travel in the lunar sunshine. When it ended, the test had simulated mileage and abuse far beyond those envisioned for an actual mission on the moon. "We had one wheel which we tested to destruction, where the wires broke in this area here," Pavlics said, pointing to the sidewall. "But it took more than two hundred miles."

24

AS GM AND BENDIX DID BATTLE THROUGH THESE YEARS OF START-and-stop NASA plans, a third company was angling for the agency's rover business. Grumman Aircraft Engineering was no stranger to government work: the Long Island, New York, firm was the prime contractor on the Apollo lunar module and had built a lot of warplanes over the previous four decades, including the bulk of the navy's carrier-based fighters during World War II. And though they never achieved much traction, the company's rover concepts were so imaginative that they deserve celebration, belated though it is.

Much of the credit for Grumman's efforts belonged to an engineer named Edward G. Markow. He was a second-generation company man. His father died in an on-the-job accident at Grumman when Ed was barely walking, and, aside from breaks for the service and college, Markow himself had worked there since high school. "There wasn't anything he loved more than his work," his wife, Margaret, told me. "I used to kid people that if he ever took his shirt off, they'd see he had a big *G* tattooed on his chest." As a structural designer he displayed a gift for unexpected ideas, and in 1957 was promoted to an in-house think tank for scaring up new products. "They let him do some crazy stuff," recalled his son, Jim. "But the space program was where my dad's heart was. It was pretty much all he would think about." When the Apollo program came along, Markow was picked to be Grumman's engineering manager for lunar exploration.

Thirty-four when he got the job, he spent much of his energy pondering how a rover should best connect with the moon's surface. Grumman experimented with tracks, then with a series of elliptical wheels, before Markow hit on his "metalastic" solution: a wheel with flat, curved spokes, arranged like the chambers of a nautilus shell, enclosed by a band of sheet metal. When not shouldering a load, the wheel was round; when supporting a vehicle, it squashed into an ellipse, with a contact patch equal to that of a rigid wheel three times its size. Markow patented the design.

This metalastic wheel was central to Grumman's 1963 studies into what eventually became Molab, and a design that would have been at home in a comic book: an articulated vehicle in two sections, the front resembling a streamlined Oscar Mayer Wienermobile, and the rear, a low, square-edged box, each on a single axle and joined by a flexible link. In the sausage-shaped cabin, the crew could travel in air-conditioned comfort. The drive system and fuel were out back. It was truer to Greg Bekker's train system than any GM design except the six-wheeled Surveyor LRV. No surprise there, because Markow knew Bekker and admired his work. Grumman designed variants to house one, two, or three astronauts, as well as an optional third unit for the train.

An appropriately spacy view of Grumman's futuristic, articulated Molab concept featuring Ed Markow's metalastic wheels. (MARKOW FAMILY)

When the Marshall Center selected the contractors for its Molab program, however, Grumman was left behind in favor of Boeing and Bendix—this, despite having built a working, Earth-gravity version of its concept. On a four-acre clearing in the scrubby pine barrens outside Calverton, New York, the company built a cratered moonscape of iron slag and cinder, and there unleashed its full-size Molab demonstrator. It handled the terrain with a grace that belied its size and quirkiness.

Ironically, the Grumman design was probably the only part of the Molab program that most civilians ever got to see. Newspaper reporters visited Calverton for rides on the oddball craft, and their stories were invariably accompanied by photos. In February 1967 venerable CBS News anchor Walter Cronkite drove the Grumman Molab on *The 21st Century,* his weekly TV program. Markow rode shotgun.

Undeterred by the Molab setback, Markow dived into designing Grumman's LSSM candidate. He would disappear into his lab for days. Even family vacations were research oriented: Markow took Margaret and the kids to Oregon's volcanoes to study the soil on their slopes. "We were checking out volcanic fields in a rental car, to see whether the area resembled the surface of the moon," his daughter, Elizabeth, remembered. "We'd be out in the middle of nowhere."

For his LSSM, Markow dreamed up a revolutionary new wheel that left his metalastic design in the dust. The cone or conoidal wheel—or Markow wheel, as it came to be called—was shaped like a bowl turned on its rim, open side out. It was made of a flexible fiberglass-epoxy blend that flattened under a load and stayed glued to rough ground while soaking up collisions with rocks and other obstacles. "I remember my dad coming out of the garage and saying, 'This is my design of the wheel that's going to go to the moon,'" Elizabeth said. "He had a cone wheel, and he said, 'See, this is flexible. When it hits a crater, it's not going to bounce. It's going to absorb it.'"

Pointed slightly downward, the wheel created a nice, big, rectangular contact area. To boost its traction in loose soil, Markow added titanium grousers to the rim's edge. Putting the design to a real-world test, he mounted the cone wheels on a Vietnam-era jeep and drove it through marsh, sand, and fields of snow. "It went over everything," said Jim Markow, who rode in it with his father. "It went over that moon area at Calverton like it was no problem at all." Markow even put about a hundred highway miles on the wheels, at speeds up to fifty miles per hour. Not only did they offer a smooth ride, they didn't seem to wear. And any soil they scooped up, they promptly flung back out.

That wheel was just one surprising element of Grumman's outlandish LSSM. Like the company's Molab prototype, this was a two-unit train, with a flexible link joining its front and rear. The wheels, out at the tips of long, shock-equipped outriggers, did not steer. On a command to turn right or left, the craft bent in the middle, and did so quickly. Its reflexes, combined with its wide stance—it was almost square in its dimensions—enabled it to handle the Calverton moonscape more like a sports car than a lunar rover, diving into craters and zooming back out without leaving the ground or losing its footing.

We know this because Grumman again built an Earth-gravity version of the machine. Driving it apparently took some getting used to: its lone occupant sat way up front, between the front wheels, with his feet in stirrups that jutted from the craft's leading edge. "When you went into a crater," Jim Markow said, "it looked like you were going to dive into the ground face-first."

Yet again the company failed to make the cut with NASA. Markow kept working on his concept anyway. He was an optimist, his family told me, and was convinced that hard work and good design would pay off. The Grumman vehicle's strengths seemed obvious to him, and the performance of the cone wheel, in particular, was indisputable. Its "strongest points," the company argued in a later report, "are its light weight, high reliability, and favorable structural characteristics. Its weight for the large diameter wheel was the lightest of those compared." It could be fashioned from aluminum, titanium, or an alloy of the two, in addition to the fiberglass-reinforced plastic that Grumman preferred.

Its only negative was its volume: the wheel took up about 15 percent more space than traditional forms. But the Grumman LSSM concept, like those of Boeing and Bendix, was on the large size and was intended to ride to the moon aboard its own lander, so that wasn't too major a drawback.

All in all, Grumman's moon cars were serious contenders, more so than their record with NASA suggests. Three decades later, the space agency was ruminating over what a next-generation combination of lunar shelters and rovers might look like. Damned if its conceptual drawings didn't depict the vehicles equipped with Markow's conoidal wheel.

25

NOT ALL OF THE ENGINEERING WORLD WAS SOLD ON GREG BEK-ker's assertion that wheels, of any kind, were the only logical choice for moon mobility. TRW, an engineering, electronics, and aerospace outfit with a long NASA history, pushed the idea of a lunar hovercraft well into the 1970s. Big, round, and noisy, its LunaGEM used rocket propellants to fill its "air" cushion and could theoretically skim across ground and over two-foot obstacles at far higher speeds than a rover with wheels. But its speed came at a price: the LunaGEM wasn't quick to turn or answer other handling commands, which spelled trouble on the pocked and boulder-strewn lunar surface. "Our present knowledge of lunar terrain tells us a significant amount of maneuvering would be required," read a Marshall Center assessment. "LunaGEM performance penalties drastically [limit] maneuvering or science stops."

More surprising was the "Lunar Worm" envisioned by the Aeronutronic Division of Philco Corp. and studied under a NASA contract in 1966. Thirty-five feet long and six to ten feet in diameter, it inched or snaked its way across the ground. Philco explored several variations on worm/snake/caterpillar motion. "The concept wherein the bottom surface of the vehicle moves in the form of a traveling wave similar to the motion used by snakes and centipedes appears to be potentially the most versatile," the company advised, claiming it "capable of moderate to high speeds over rough terrain with a smoother ride and less vehicle bounce" than traditional vehicles. It also promised "good maneuverability, good propulsive efficiency," and a "better ability to climb obstacles and to bridge crevices." This worm could be used on an array of surfaces, "including even fluid-like soils through which it is capable of swimming."

Philco also researched an "extension-contraction bellows concept, wherein the vehicle alternately shortens and lengthens." It was a simpler machine, but required "the most power because of friction in sliding over the terrain." When it came to moving two astronauts long distances, Philco recommended its traveling wave approach. Such a worm might be up for a one-hundred-mile trip at an average speed of five miles per hour, and would be fairly comfortable, with a cabin ten feet long. One drawback: it would weigh a ton. Literally.

NASA took a pass. The worms went nowhere.

A still more startling approach to moon mobility belonged to the "Lunar Leaper." This two-man pogo stick called to mind Hermann Oberth's moon car of 1959 and could, according to its inventor, bounce over the lurrain at twenty-plus miles per hour. The Leaper consisted of two pods, one a crew cabin and the other the powerplant, built to slide around a forty-foot, hollow pole on a cushion of compressed gas. The crew would board the craft when the pods were sitting on the ground, at the pole's bottom; when activated, gas would shoot the pods thirty feet up the pole, where they locked into place as the gizmo launched into a hop two hundred feet high and four hundred long.

Halfway through this fifteen-second flight, the lower end of the pole would swing forward in preparation for landing. At the pole's bottom was a flexible, cleated foot, to cushion the resulting impact; the pods would slide back down the pole, and momentum would pitch the contraption forward, ready for the next hop. Gyroscopes would keep it balanced at the appropriate angles.

The Leaper's champion was no crackpot. Howard S. Seifert had been a research physicist in the private sector, a rocketeer at the Jet Propulsion Laboratory, and president of the American Rocket Society, and when he unveiled the idea early in 1967, he was teaching at Stanford University. Seifert argued that the Leaper offered advantages beyond speed: while airborne, the crew would have a commanding view of the surrounding moonscape and would pick its next landing spot and overall route with the help of a computer and a sighting device. Credentials or no, Seifert didn't win many backers at NASA.

That was not the case with an idea that, in retrospect, seems all the more terrifying for the genuine respect it got from the space agency: a Lunar Flying Unit, a small, rocket-powered rover that would zip an astronaut over the lunar surface at seventy to two hundred miles per hour, depending on which document you believe. The LFU would run on leftover fuel from the lunar module, which the astronauts would transfer into the flyer via hoses. Seeing as how the module's engine relied on an "ignite on contact" chemical reaction, this was in itself an accident waiting to happen. Once gassed up, the flyer might range far and fast from the lander to explore, visit interesting formations, collect samples, and so on.

Not since Icarus has a scheme for manned flight been so fraught with peril, but that didn't stop the Manned Spacecraft Center in Houston from chasing it for years. The LFU traced its conceptual origins to a rocket-propelled backpack designed and built by Bell Aerosystems in the fifties and early sixties, partly under contract to the U.S. Army. The Bell rocket belt, as the company dubbed it, ran on hydrogen peroxide, but not for long—the small amount of fuel its wearer could carry limited flights to twenty-one seconds. That was not its only shortcoming: it weighed more than one hundred pounds, was almost suicidally difficult to control, and required its operator to wear fireproof pants. The army walked away.

But not NASA. It reasoned that absent the earth's gravity and atmosphere, a small rocket wouldn't have to work so hard, and would thus sip fuel rather than chug it. By 1967, the agency and its contractors had "established the feasibility of a variety of flying vehicles," according to a draft LFU program plan. The draft envisioned a flyer weighing two hundred pounds empty, and capable of carrying up to three hundred pounds of astronaut and gear. It would have simple, all-mechanical flight controls and a bare-bones guidance system, and it would be compact enough to stow in the lunar module's descent stage. Houston hoped to have it ready by 1971.

As of October 1968, the Manned Spacecraft Center's Lunar Missions Office had received concepts from Bell Aerosystems, as well as North American Rockwell and TRW, and had passed them around for comment. The following month, a draft NASA document soliciting proposals for a flyer training simulator demonstrated that the more the agency talked about the idea, the scarier it got. "Two vehicles are considered necessary to effect rescue if difficulties develop

The U.S. Geological Survey's Gene Shoemaker modeling a Bell Aerosystems rocket belt during a 1966 demonstration outside Flagstaff. Thankfully, he didn't try to fly the thing. (USGS)

with one of the units," it read, not broaching what would happen if the second LFU ran into trouble. "The flyer will be used for a number of different types of sorties. . . . In one case, the astronaut might make a large loop covering a total distance of 25 miles without any intermediate landings before returning to the LM. In other cases, the astronaut might fly five to seven miles from the LM, land in a (canyon) or on a dome, explore the area for an hour or two, and return to the LM base. On still other sorties, he might make four or five landings, all no further than one mile away from the LM."

In short, an astronaut flying one of these things would be alone, and perhaps a long way from his partner. Compounding that risk were the formidable skills necessary to fly. "To utilize propellants efficiently, the pilot would have to pitch forward up to 45 degrees for acceleration and pitch backward up to 70 degrees to effect braking and may fly at velocities up to 300 feet per second," the draft read. "The pilot will fully control the flight attitude, velocities, and altitude of the vehicle while navigating to his targeted destination, and in some cases, making observations or activating cameras and instruments during flight, all with a minimum of electronic or mechanical aids to guide him in these tasks."

Astronauts found it difficult to perform simple chores in their pressurized space suits, let alone multitask while wrestling the controls of a rocket-powered death trap. And the LFU left zero margin for error. "The moon is a very poor place to gain experience in the use of the flying vehicle," as the draft said, "because if difficulties do develop very little can be done to assist the pilot in overcoming his problems."

The Manned Spacecraft Center forged ahead. In January 1969 it awarded contracts for detailed studies of scaled-down flyers, with just one hundred pounds of payload, to Bell Aerosystems and North American. The companies dreamed up very different approaches. Bell proposed a twin-engine flying platform, which the standing astronaut would maneuver with joysticks similar to the controls in the lunar module. It would fly at speeds up to one hundred feet per second, or just shy of seventy miles per hour. The astronaut's knees would absorb the forces of landing, because the craft lacked a suspension of its own.

North American's machine was bigger and more complex—a four-engine flyer in which the astronaut was strapped into a recumbent seat—and, while ostensibly safer, more of a hassle. Because the engines were slung low beneath the pilot's chair, the company worried that the blast of takeoff might fling shrapnel, endangering all and sundry. The crew thus had to drag the empty, 304-pound LFU forty feet from the lander, and center it on a square of fabric before breaking out the fuel hoses. When the flyer landed out in the field, its pilot would cut the engines a few feet from the ground, allowing the shock-absorbing legs to soften the resulting bump. Assuming he survived, he'd prepare for his subsequent takeoff by unrolling another fabric square. Although the fueled LFU would weigh a lot more than the empty craft, and this time the pilot would be working alone, he'd then have to drag his flying machine over it.

With the smaller payloads specified by NASA, neither concept was capable of carrying a second astronaut. If a Bell flyer quit working far from base, a second craft could fly out to deliver life-saving air and water for the long walk back. If, on the other hand, the North American LFU ran into trouble, no rescue was coming, as only one would fit on the lunar module. Consequently, that machine wouldn't be allowed to fly more than 4.6 miles from base.

With those concepts, the lunar flying unit hit the ground for good. The companies presented their ideas to Houston and NASA headquarters officials a month before the first moon landing, and NASA—an agency renowned for its caution, for its insistence on redundancy, for its penchant for testing and testing again—finally, thankfully, let the half-baked notion die.

Years later, a jetpack-style device was supplied to astronauts on the Space Shuttle, for use when performing untethered space walks outside the orbiter. But an older and wiser NASA had the good sense to retire it as an unnecessary risk after just three missions, and to reserve its replacement for emergencies.

PART FOUR

"WE
MUST
DO THIS!"

26

EAST OF FLAGSTAFF, ARIZONA, THE PONDEROSA PINES FALL AWAY from alongside Interstate 40 and the rolling Colorado Plateau stretches to the horizon, bare but for rocky sand and chaparral. Thirty minutes out of town, a rise appears off to the south: a low, snaggle-toothed ridge jumbled with boulders. Coon Butte, it used to be called. Climb up top, and there comes the revelation that this is no butte; its far side is a chasm in the earth three-quarters of a mile across and 560 feet deep. The uplift is the rim of a gargantuan crater.

From the early days of its European settlement, the surrounding desert was known for the small meteorites that cowboys and prospectors found among its creosote bushes and scorpion weed. In the late nineteenth century, the American geologist Grove Karl Gilbert, who had figured out that the moon's craters were more likely formed by impact than by volcanism, examined Coon Butte and decided that here the opposite was true. When he measured the crater's debris on the surrounding plain, he found that it would fill the hole, with no room left for a meteorite he reckoned had to be the size of an office building. Besides, a volcanic field around Flagstaff, forty miles away, signaled that forces churning deep inside the planet rose close to the surface here. The scattered meteorites had to be coincidence, not evidence.

Gilbert's judgment became the scientific consensus. Then in 1902 a lawyer, adventurer, and mining entrepreneur named Daniel Barringer heard about Coon Butte and bought it sight unseen, confident that Gilbert was wrong—and that under the crater's floor rested a meteorite loaded with enough iron, nickel, platinum, and iridium to make him rich. Barringer began mining the crater's floor in 1903, finding just enough trace of the metals to keep at it, year after year, to the brink of financial ruin. Through it all, he insisted his crater was a gift from outer space. In the meantime, people stopped calling it Coon Butte. The nearest post office was in the desert outpost of Meteor. The crater took its name.

Why I am telling you all this? Because Meteor Crater had a formative role in the creation of a government agency, and that agency had a hand in what the Apollo crews did on the moon, and how they did it—and, eventually, in the success of the lunar rover.

Let's skip ahead a few years to 1952, when a geologist named Eugene M. Shoemaker, twenty-three years old and four years into his service with the U.S. Geological Survey, was assigned to its search for uranium on the Colorado Plateau, and wandered to Meteor Crater. Captivated, he returned every chance he got. A while later, on a visit to the Nevada Test Site north of Las Vegas, he looked into craters scooped by nuclear blasts and was struck by their similarity to the much larger hole back in Arizona. In an instant, intuitive leap, Shoemaker understood why Karl Gilbert and Daniel Barringer found so little of a meteorite there: it had hit with such force that it vaporized. He would go on to prove it.

Shoemaker's fixation on craters got him thinking about the moon. On a back-road drive one morning, he would say later, he saw it hanging pale in the sky and suddenly felt compelled to go there. This was no idle wish; it seemed destiny. The Mercury Seven had not yet been chosen. America had no announced plans to hire astronauts. All the same, Shoemaker resolved to be one.

His first step was to interest the Geological Survey in studying lunar geology. His bosses were cool to the idea, so Shoemaker pursued the study himself, before and after the agency abandoned its uranium project and transferred him to its offices in Menlo Park, California. Soon he invented a term for what he was doing—astrogeology—and, although he got a lot of teasing from his colleagues, came up with the first procedures for mapping the moon's mineral makeup. In 1960, when, as his colleague Don Wilhelm put it, the Survey "had too little money and too many geologists, whereas the reverse seemed to be true at NASA," the space agency underwrote the creation of an Astrogeologic Studies Unit at Menlo Park, with Shoemaker its head. Two years later, in the wake of JFK's challenge to the nation, his tiny squad expanded into a full-fledged branch of the Survey. In 1963 he won permission to move the operation to Flagstaff.

The town already had a spacy reputation. Its clear skies and seven-thousand-foot elevation had attracted two observatories; the air force was mapping the moon for NASA at one of them. Its geology was a strong draw, too: On Flagstaff's northern fringe was a cluster of extinct volcanos that included the state's highest mountain, as well as hundreds of lava flows and cinder cones, or small volcanic vents. The Grand Canyon was an hour and change to the north. And, of course, there was Meteor Crater.

As Shoemaker arranged the move, he learned he had Addison's disease, an adrenal gland malady that erased any chance that he'd qualify as an astronaut. He responded to the news by throwing his outfit behind the men who would. Some of his scientists judged the Surveyor LRVs and evaluated the mobility test articles at the Yuma Proving Ground. Others worked with Marshall Center engineers on lunar soil studies or helped plan and execute the Ranger and Surveyor programs. The branch's geologic maps of the moon helped pinpoint the most interesting and scientifically productive spots for the manned landings to come.

But its greatest contribution, by far, was in training. Between 1963 and 1972, the Survey organized about two hundred geologic field trips for Apollo astronauts—to the Rio Grande Gorge in New Mexico, the Nevada Test Site's nuclear craters, the California desert, Hawaiian volcanoes, and the treeless sweeps of Iceland. Many of the excursions were to sites around Flagstaff, especially the volcanic field. The most celebrated of its blowholes is Sunset Crater, which erupted in about A.D. 1085, spewing the Bonito Lava Flow to the crater's northwest and raining cinders on more than eight hundred square miles of the surrounding country. South of the crater is Cinder Lake, a flat expanse of dark, gray-brown cinder, much of it pea sized, and in places layered fifty-five feet deep. To the Geological Survey, the Sunset Crater area seemed an ideal lunar analog, and Cinder Lake especially so.

Or it would be, with some modification. In July 1967 USGS crews dug forty-seven holes in the field, stuffed them with dynamite and fertilizer, and blew them into craters ranging from five feet to forty-three feet in diameter. When the dust and cinders settled, the holes were a match in size and placement for those on the moon's Sea of Tranquility—a very specific area, five hundred feet square, that the Survey itself had recommended for the first Apollo landing. Three months later, the team added ninety-six craters to the original forty-seven, expanding the field to eight hundred feet on a side.

The match was uncanny. Compare photographs of the moon that the geologists used as a reference with aerials of the Arizona site, and it's striking how close the duplication came. Stand in the field's middle, and it's even more convincing. Treeless, its dark expanse interrupted only by the odd clump of chaparral, and with Sunset Crater and a cluster of cinder cones rising to the north, what came to be called Cinder Lake Crater Field No. 1 is a lonesome, otherworldly place.

I ventured there in October 2019. With me in a rented SUV was Laszlo Kestay, a planetary volcanologist and until recently director of the Survey's Astrogeology Science Center in Flagstaff. "Definitely bring hiking boots," he'd told me on the phone a few days before. "The cinders are technically broken glass. It's a very poor-quality glass, but it's a glass. You don't want any getting down in your shoes." Kestay, a compact and fit fifty-one, had a mop of jet-black hair and a boyish enthusiasm for anything related to the Apollo program. "I was born in 1968," he said. "When I was little, I can remember watching TV and watching the astronauts with the rover. Those are the earliest memories I have, just about.

"This place we're going—as soon as I saw it, I wanted to bring people to see it."

The first of three crater fields the Geological Survey created as lunar analogs in the wilds around Flagstaff, photographed shortly after the dust settled. (USGS)

North of town, we turned off U.S. 89 onto a narrow two-laner that doglegged through pines and skirted the edge of a subdivision of sun-bleached ranchettes. About a mile from the highway, Kestay pointed me off the blacktop and onto a sandy path snaking through the trees. We emerged from the woods beside an immense landfill. A half mile down its edge, just as I was beginning to wonder about our inauspicious setting, Kestay directed me away from the dump to fishtail across a widening plain of cinder. It grumbled loud under the tires for the three minutes it took to reach a low wire fence. A small sign hung from one of its strands:

During the late 1960s and early '70s, the area behind this fence was used to train astronauts for the Apollo Space Program. Craters were blasted out with explosives and rocks brought in to simulate the surface of the moon. Here, astronauts learned to walk in space suits and use equipment to prepare for lunar missions. Similar areas were created in Texas and Florida, but no longer exist.

This is the only remaining astronaut training area. Please help us protect this important part of American History.

We were the only people around. We left the car, passed through a break in the fence, and set off across the field, the cinders light and shifting readily under our boots, crunching loudly with each of our steps. It felt and sounded as I imagined it would to cross a sea of Cocoa Puffs. "They really tried to give a good sense of what being on the moon was like," Kestay said. "This surface isn't exactly the lunar regolith, but it gives you some idea. You're not walking on concrete. You're not walking on sand. It's something in between."

I had studied aerials of the field and knew that we walked among craters, but looking around, could see none of them; Cinder Lake was a seamless charcoal plane. "One of the main lessons we learned here, just as a practical thing, was that when you look at a picture taken from overhead, you see all the craters on the moon," Kestay said. "They're just obvious." He raised his voice over the sound of our crunching. "When you're walking on the surface, you don't see them." We paused and took in the field around us. "This was a problem, particularly once they started driving," he said. "You go too fast, you don't see a crater in time, and just slam right into it."

We arrived at an earthen ramp piled into the field's center. "Come on up here," Kestay said. "As you start coming up, you start seeing all the craters." Halfway up, craters materialized all around us. The ramp, Kestay explained, was built to the height of the lunar module's descent stage—its lower half. During training sessions, the Survey trucked a plywood mock-up of the ascent stage onto the ramp, so that the view from its cabin simulated what the astronauts could expect to see in their first minutes on the moon. "That height was a good one to make the first observations after they'd landed," he said. "Land, shut down the engines, and make a description of what you saw—that's what they were supposed to do—and they would rehearse that process here. They'd practice making that description, pretend they were climbing out, then start practicing EVAs." Short for "extravehicular activities," NASA-speak for carefully planned walkabouts and drives.

I did a slow turn to take in the panorama of craters, large and small, that stretched off in every direction. "The astronauts really learned a lot from doing the exercises here," Kestay continued. "How do you use a rock hammer? In one of those suits, it's nearly impossible. So they had to figure out how to do basic things like that. And once they started training to drive on the moon, coming out here was really important."

We descended the ramp and headed into the field, stopping at the edge of one crater, then another. Time and weather had softened their edges. Cinders had washed down their sides to bury their floors. A gusty breeze swept over the field, and clumps of sage and creosote, blooming bright yellow, quivered at their rims. The biggest crater lay at the field's edge. As we trudged toward it, we passed a bright white rock resting on top of the cinders, then another chunk of light-colored stone that seemed out of place. "Sandstone," Kestay said. "This particular patch was used as a final test of the astronauts' geological understanding. The scientists seeded it with all sorts of rocks that put together a story that the astronauts were supposed to sort out, to figure out what was going on."

That simple ploy was so successful that the Geological Survey later shipped two railroad freight cars of basaltic cinders from Arizona, a dump truck's load of granite and gneiss from the Texas Hill Country, and a pile of crystal anorthosite to a smaller moonscape at the Kennedy Space Center, to give astronauts a last dose of geologic training before they flew.

We reached our destination, a crater twice as wide and deep as the others. On its floor were the ruins of a campfire and a scattering of empty beer bottles. "This is the shape of the real lunar craters," Kestay said, "and you can see why they didn't want the astronauts to go into them on the moon. If you went in, you wouldn't get out." The hole was not more than a dozen feet deep, but its sides were sharply angled. It looked like it would take effort to climb out of it in jeans and hiking boots, let alone an extravehicular mobility unit.

We slogged back to the SUV and started for a second field. A year after they created the first, Survey staffers burned away the chaparral from a much bigger piece of Cinder Lake a mile to the north, and set off 426 explosions to create Cinder Lake Crater Field No. 2. Here they took even more care to replicate lunar conditions. They knew that when a meteorite strikes the moon, it creates a spray of ejecta that radiates outward from the point of impact. A meteorite crashing down later sends out rays of debris that layer atop those of the first. That being so, a young crater is easy to pick out, for its ejecta lies on the surface, while the debris fields of old craters are buried deep. So the Survey started July 27, 1968, by touching off dynamite and fertilizer to create 354 old craters. Later in the day, they set off sixty-one blasts to create middle-aged craters, and finished by blowing the cinders out of eleven young ones. The result was a training ground of overlapping craters and ejecta rays.

Crater Field No. 2 was much easier to reach: a proper road took us right to its edge. In the years since its Apollo heyday, it had been used as an off-road recreation area and was crisscrossed with four-wheeler tracks. "There was a fence up early on, but it got knocked down and people drove right over it," Kestay said. He nodded toward a lonely pine a quarter mile off. "That tree out there is the edge of the crater field." The night before, the temperature had dropped below freezing in Flagstaff. Now the wind had stilled, and the sun, intensely bright, was directly overhead. Our trek across the cinders was slow, sweaty business. The only other human presence was a Honda Accord, loaded with teenagers, struggling bravely across the uncertain ground far to our west.

Out in the field's middle, once-deep craters had eroded to shallow bowls. Some had been all but erased by all-terrain vehicles, which had chewed up their sloping rims until little slope remained. "They've messed it up so much," Kestay sighed. "That was a crater over there." He pointed to a slight depression I would have missed had he not called it to my attention. "There's another one over there. This is always changing.

"Still, you can see why the astronauts couldn't drive too fast. And you get a good sense of the many different sizes of craters they might run into. Some of these are still big enough that if you brought a regular four-wheel-drive out here and drove into one, you'd be in trouble."

Among the deeper holes were faint cavities that might have been craters or might have been excavated by spinning tires. It was hard to tell. Still, I was grateful this place had survived, even in its ravaged state. The Survey built a third field an hour south of Flagstaff in February 1970, after it became clear that snow and northern Arizona's winter cold rendered these first two unusable for several months a year. Again using dynamite, it blasted 362 overlapping craters into thirty-five acres of desert hardpan. Later in the year, fourteen new craters were added. Black Canyon, as the training ground was known, has disappeared without a trace. A subdivision occupies the spot today.

"What an eerie, wonderful time capsule," I said. "Fast disappearing, but still."

Kestay nodded, looking a bit glum. "It is. And unfortunately, I don't see any practical way to freeze it. It's going to keep wearing down. Even without the four-wheelers, it's been more than fifty years, and things happen."

Heat radiated from the cinders at our feet. I looked out over the empty expanse, struggling to picture it busy with scientists. It saw heavy use in the 1960s: even when the astronauts weren't around, Cinder Lake helped refine the procedures they would use on the job. The Geological Survey's people donned bulky space suits and worked out how to best gather samples while wearing them, an effort that produced some of the rakes, scoops, and other tools the moonwalkers used. Insights from these outings also informed the detailed map packages the astronauts carried on their lunar traverses. Rehearsals on the cinders helped perfect communication between explorers, scientists, and Mission Control, too.

"One of our mainline efforts here was to try to work out the most efficient way that you could do human exploration," Shoemaker explained years later. "My notion, and this was the thing in the back of my mind all the way along, was . . . to really try to get a maximum return on what the astronauts themselves could do on the moon."

We started back to the SUV, crunching. "Some people have claimed that geology is actually a historical science, and not a physical science," Kestay was saying. "And they're halfway right, because you're really looking back through time. You look at a landscape and try to see, through time, all the things that happened in that place." That was one of the reasons for Shoemaker's interest in the moon. On Earth, the distant past can be hard to read. The forces of erosion, glaciation, volcanism, plate tectonics—they all conspire to obscure the nature of a given spot over time. The moon, however, has been dead since infancy. Its rocks are snapshots taken billions of years ago, and, because they're part of what may well be the least-changed major body in the solar system, they testify not only to its own history, but to ours. As von Braun had written in *Collier's*, lunar exploration could help answer our questions about the origins of the earth and the formation of the universe.

Science had been a stated aim of the lunar program from the start, but no *one* science had been identified as its focus. "You had astronomers and physicists who thought it would be great to be able to make observations from up there," Kestay said. "Shoemaker said, 'No. We're doing geology.' And there are places in history where a single individual's force of will tips things." By August 1967, when a NASA-sponsored scientific summit convened at the University of California at Santa Cruz, Shoemaker's view prevailed. The conference held that, while "flight-operations ability is, without question, the primary criterion" in selecting crews for lunar missions, the "next most important factor" should be "ability in field geology."

We paused for a last look around. Craters we had skirted just minutes before were gone. The field was as flat and unvarying as a lake on a breezeless day. "We have the same problem out here," Kestay said, following my gaze. "It's really hard to see the craters." We resumed our hike. "Have you been out to Meteor Crater?"

I told him I planned to drive there the next day. "Well, you see the same thing there, on a grander scale," he said. "If you know what to look for, you can tell there's a hole there, but there's not much of a clue."

27

THE SUMMER STUDY OF LUNAR SCIENCE AND EXPLORATION, AS that 1967 gathering in Santa Cruz was formally known, was unequivocal about another priority. "The most important recommendation of the conference relates to lunar surface mobility," the attending scientists wrote. "On the early Apollo missions, it is expected that an astronaut will have an operating radius on foot of approximately 500 meters. It is imperative that this radius be increased to more than 10 kilometers as soon as possible." The scientists suggested that NASA waste no time designing a lunar flyer and—perhaps even more urgently—plan for a Saturn V dual launch and pack an LSSM aboard the space truck.

Bear in mind, this was two years *before* Apollo 11, yet the conference report didn't hedge any bets about the lunar program's success. Its authors treated a successful first landing as a done deal, expressing impatience to get it and other early, adventure-mode missions out of the way so that the astronauts could get down to serious science. And their support of the flyer notwithstanding, they recognized that the astronauts wouldn't get much done on foot.

Alas, by the time these findings were committed to print, they'd been leapfrogged by events. The first eight months of 1967 had already been tough for NASA. In January the agency had been devastated by tragedy when its first three Apollo astronauts—Mercury and Gemini veteran Virgil "Gus" Grissom, Gemini veteran Ed White, and rookie Roger Chaffee—were killed by a flash fire that ripped through their command module during a routine test at the Kennedy Space Center, weeks before they were due to launch. The Apollo 1 fire knocked NASA back on its heels, raised public doubts about its competence, and sparked a lengthy investigation within its own ranks and inquiries in both houses of Congress. The command module required a redesign. Manned flights were put on hold.

Now came the kicker: Congress slashed NASA's budget for fiscal 1968—sparing the immediate Apollo moon effort but gouging deep into the agency's plans for what would follow. The cutback scrapped any prospect of two-rocket missions, at least for the foreseeable future, and without those, there would be no LSSM. "Hence, the plans are optimistic in outlook," Homer E. Newell Jr., NASA's associate administrator, wrote in a preface to the Summer Study's report, "and exceed the capability of the agency to execute."

Headquarters sent word to the Marshall Center to halt procurement for any further lunar vehicle studies. In October the existing LSSM contracts at Boeing and Bendix were allowed to expire.

28

IN SANTA BARBARA, THE GENERAL MOTORS TEAM THAT HAD worked on the LSSM was scattered to other projects. Sam Romano and Frank Pavlics were not inclined to quit work on a rover, whether or not NASA was underwriting it—the pair had devoted too much time, energy, and brainpower to the effort to simply walk away. The budget situation did require a shift in thinking, however. Without a second Saturn to carry cargo, any rover bound for the moon would have to be carried on the same lunar module that carried the crew. Could it be done?

Unknown to Pavlics, engineers at Bendix and Grumman were mulling the same question and facing the same challenges. The lunar module was actually two spacecraft in one, and defied every notion of how a flying machine should look—it called to mind a doorknob balanced on a plant stand, all of it wrapped in foil and bristling with antennas. Its roughly square descent stage, spider's legs jutting from its corners, was topped by a faceted, asymmetrical upper stage containing the crew cabin. It was designed to make a soft landing on the moon's surface using the engine in its lower half, and leave that piece behind at mission's end, the astronauts blasting back into space in its bizarre upstairs.

Peculiar as it was, it was hard to identify any storage space at all on the thing. But the module was scheduled to receive an upgrade after the first few moon landings, and this improved version, designed for a three-day lunar stay, promised to have a greater capacity for payload. On trips to Grumman and NASA, Romano learned that one of the wedge-shaped cavities between the descent stage's legs might be available. This "quadrant" measured sixty-nine inches wide, by sixty-four inches tall, by forty-seven inches deep, with the point of the wedge facing inward—little more than the cargo volume of a full-size station wagon. About the size and shape of a pup tent standing on its end.

Romano delivered this news to Pavlics, who wondered how to shoehorn a vehicle into so tiny a space. He reached one conclusion quickly: the six-wheeled style of moon car that had been GM's hallmark wasn't going to fit. At most, he might be able to pull off a stripped, downsized version of the LSSM's front four-wheel section, and even that would involve some complicated origami. Not only would the machine have to fold into a tight package, it would have to assume the same wedge shape as the hole. Pavlics started with an assumption: the chassis, once folded, could be no bigger along its largest plane than the roughly five-by-five-foot mouth of the storage bay. He played with the ways a vehicle a bit smaller than the LSSM, the chassis of which wouldn't need to be more than five feet wide, might fold into a rough square.

His solution: Both the front and rear pieces of the chassis—the parts that included the wheels—could fold back onto the middle. And the suspension could be hinged so that each wheel assembly then folded upward and inward, over the folded chassis, at a forty-five-degree angle. The wheels would meet in a point corresponding to the interior of the wedge.

As Romano and Pavlics later told it, it took an intense four months of trial and error to arrive at this answer. To test his thinking, Pavlics built a one-sixth-scale model of the folding chassis and constructed a mock lunar module bay to go with it. It fit. Even better, the stored model's belly—the underside of its center section—faced outward, shielding the rest of the vehicle from the rigors of space. Working with the model helped Pavlics figure out how the rover might unfold from the lunar module—its center section first, lowering like a drawbridge; its rear section flipping over; back wheels swinging up to lock in place; those wheels dropping to the ground as the front chassis unfolded. Satisfied that it was ready to show to NASA, Romano and Pavlics took the model on the road—but outfitted it first with an electric motor and radio remote control, and made a short film depicting its deployment.

Pavlics still has the model. He keeps it in a closet at home under a protective Plexiglass shroud. "I built it with my hands," he told me, as he placed it on his dining table and removed the cover. "Some of it we machined at the lab, but mostly I built it here. My wife helped." It was faithful to an actual Santa Barbara product, with wheels fashioned from stainless steel window screen, folding seats, and, as a final flourish, a G.I. Joe action figure in a silver Gemini space suit, wearing a to-scale Apollo-style personal life support system backpack. He showed off its various components: four-wheel independent suspension, true

Frank Pavlics shows off his folding, one-sixth-scale model of the rover at his Santa Barbara home, June 2019. (EARL SWIFT)

to the actual rover; steering linkages that mimicked a real car's; tiny hooped bump stops inside the wheels.

He folded up the model—front section first, then rear, then the wheels, one by one. "When it is folded up like this, it can fit into the triangular space," he said. I watched as he unfolded it, locking its wheels into place with care, and replaced the G.I. Joe in the driver's seat. "My son Peter's G.I. Joe," he said. "I took it away from him." He looked at me and smirked. "He's still waiting to get it back."

Fifty-plus years ago, Pavlics and Romano toted the model into the Marshall Center's Building 4200. "I gave a presentation to the technical people," Pavlics told me, "and they really liked the idea, and they said, 'Hey, why don't you show it to Wernher von Braun?' So they took us to his office." A Marshall engineer named Lynn Bradford led them to the ninth floor, knocked on the director's door, and opened it just wide enough for the model. Using his radio handset, Pavlics steered the rover into the room and across the carpet toward von Braun's desk.

He was talking on the phone when the rover caught his eye. "He couldn't see anybody, just this thing coming at him," Pavlics said. "He put down the phone and said, 'What is *this*?' We walked in and explained what it's all about. We showed him how to fold it and how it would fit into the lander."

Von Braun was impressed. When they finished, he slammed his hand on his desktop and said, "We must do this!"

29

IN THE LORE SURROUNDING THE ROVER, THAT EXCHANGE IS USU-ally framed as the moment that a toy, more or less, helped change history. Marshall Center old-timers, who as a rule are not prone to romanticizing the process of developing space hardware, still talk about Frank Pavlics and his model. Among them: Sonny Morea. "He's the guy who made the sales pitch that made it happen," Morea told me. "That meeting was the spark that made people say, 'Let's look at this seriously. Let's put out a proposal and see if we can do it.' There's no question about it."

Would the rover have come along regardless? Perhaps. It's worth remembering that GM wasn't operating in a vacuum. Bendix and Grumman persevered on their own concepts, and there's every reason to think they were discussing them with NASA. Besides which, the GM visit occurred in the midst of a paradigm shift within the space agency: As money tightened, and the prospects for post-Apollo moon visits became ever more doubtful, NASA's leadership was grappling with the idea that it would have to make the most of the few missions it had left. A rover offered a lot of bang for relatively little buck.

But timing was key. Had the impetus to get going on a rover happened just a few months later, it might have arrived too late for Apollo. What can be said for certain is that von Braun wielded great influence within his agency, and while a decision on whether to pack a car along on Apollo missions was not his to make, his advocacy would have been difficult to ignore. By the early summer of 1968, headquarters was again talking about moon cars. "It is now clear that a surface-mobility capability is essential for continued lunar exploration beyond the early Apollo landings," Benjamin Milwitzky of the Apollo Lunar Exploration Office wrote in a June 6 letter. He went on to say that he was forming and chairing a "Working Group on Lunar Roving Vehicles."

The rover he had in mind, Milwitzky said, would be light enough for the lunar module—750 to 1,000 pounds, say—but expandable, too, on the chance that another means of getting it to the moon came along. "The final output of the Working Group," he wrote, "should be a complete package of all the essential elements required for top-management approval of a roving-vehicle and payload-development program." The group met in Washington for the first time late that month. That October, it visited Cinder Lake, where the Geological Survey simulated a rover mission using a stand-in fashioned from a four-by-four truck chassis. The demonstration was convincing.

Not long after, Marshall issued a "statement of work" spelling out the agency's requirements for a study of small rovers that could be carried aboard the lunar module, as GM had demonstrated could be done. But the November 15 document had a twist: the six-month study would focus on dual-mode rovers—machines that could be used on the moon by the astronauts, then switched to remote control from Houston once the explorers flew home. Bendix and Grumman, which had already spent time working on the dual-mode idea, submitted proposals for the study. So did GM.

Bendix was probably furthest along in its thinking. It had in mind a small rover that would land aboard the lunar module and "be unloaded, checked out, and placed in operation by the crew," making "at least two sorties per day during the crew stay time, giving them extended exploration range out to 5 km and beyond." Before blasting off for home, the astronauts would "place the vehicle in operation under remote control from an earth control station." Guided by a TV camera, its batteries recharged with a solar panel, it would travel "on the order of 1,000 kilometers"—620 miles—to a future planned landing site, collecting samples on the ten- to twelve-month journey.

"Upon reaching the preselected landing area, the vehicle would be controlled to use its TV system in performing a site survey to determine the best landing site in the area," Bendix suggested. "After the manned landing, the crew would pick up the samples collected in the unmanned mode and transfer them to the LM for earth return. They would use the vehicle during their stay and, if the vehicle was still in operating condition, place it on another unmanned traverse upon their departure."

That'll get you thinking about what-ifs.

Grumman's concept was typically imaginative. It proposed a smaller version of its articulated LSSM, with a third unit that hooked on to the rear. The unit would contain the electronics and stereo TV cameras for remote control from Earth; once attached, this rear unit would become the automated rover's front. GM offered up several possible variants, including a one-man four-by-four and a one-man six-wheeler; the latter's rear, two-wheeled unit would have to be stored in a second quadrant on the lander.

Unfortunately, the concept was doomed from the start. A dual-mode rover was a complex machine, no matter who built it, several magnitudes more so than the simple folding buggy Frank Pavlics had shown to von Braun. Developing the hardware to enable it to operate remotely, and at a weight that could be carried safely aboard the lunar module, would require time and money. And while NASA was sensitive about both, it was time that really did in the concept. NASA launched a manned Saturn V for the first time with Apollo 8 in December 1968, just weeks after it asked for proposals for the dual-mode study. Yet already the program's end was coming into focus, and the agency quickly realized that if it wanted a rover to meet its schedule, it had to think more simply and save remote operation for later. It had to forgo contracting for an early-stage design study, too, and skip right to contracting for an actual vehicle.

30

WITH A MORE MODEST AND ACHIEVABLE GOAL IN MIND, NASA started to move fast. In the new year, a growing number of executives and work groups within the Marshall Center and headquarters focused on rovers, and the agency consulted with the Geological Survey and the Jet Propulsion Laboratory on what they thought such a machine should do. Work started on documents laying out the performance expectations for the vehicle. In April 1969 the subject came to NASA's Management Council, the agency's senior advisory body, comprising its top headquarters executives and center directors. The panel needed more information before it signed off on a rover project—not least, what it would cost.

Marshall Center officials asked Bendix, Grumman, and their own people to ballpark what it might cost to build small, stripped-down rovers. The best bet was $23 million for research and development, and $2.5 million for each vehicle, or about $30 million for three. NASA extrapolated that into a maximum of $40 million for four rovers, one for each of the last four Apollo missions, 17 through 20.

On May 23, 1969, headquarters telegrammed von Braun that he was "authorized to proceed with the development and procurement of the rover concept." The first of four machines was due by April 1971. The deadline left the Marshall Center with just twenty-two months to produce a vehicle that, at that point, was little more than engineers' doodlings. That sort of concept-to-hardware effort typically took NASA three or four years, minimum.

A few days later, the agency's Washington leadership offered up a few more details of what it had in mind. The rover was to seat two astronauts. It had to fit in the lunar module's quadrant 1, the cargo bay to the right of the ladder. It had to be simple, rugged, and reliable. Most importantly, it could weigh no more than four hundred pounds. Four hundred pounds *on Earth*, that is—sixty-seven pounds on the moon—a seemingly impossible number, even by the austere standards of the Specified LSSM. But so said headquarters, so the center would have to make it happen. One man, in particular.

Sonny Morea had not had a real vacation in years when, in late May, he flew his family to Montreal for a few days. He had worked for nearly a decade in propulsion. As deputy manager of the Marshall Center's Engine Program Office, he'd overseen development of the pandemonic F1 engine and later troubleshot the J2 that propelled the Saturn's second and third stages.

Both jobs were challenging. The blast from the F1's business end created an unholy 1.5 million pounds of thrust, a power so extreme that the engine spent its less than three minutes of duty at the very limits of control. Taming it—especially an instability problem that led to several titanic explosions on the test stands—took him more than seven years. The more civilized J2 didn't inspire nearly so much existential terror but was far more complicated, in that it had to be shut down and restarted in orbit. The restart wasn't reliable. It took two and a half years to fix.

The job consumed him. Morea's days at the office often lasted twelve hours. Rare was the weekend he didn't lug home a fat satchel of

Saverio "Sonny" Morea in the 1960s. (NASA)

paper. He was called on to travel, a lot, and it was wearing. He drank gallons of black coffee. Staying true to the Apollo program's schedule meant observing one of von Braun's mantras: "Late to bed, early to rise, work like hell, and advertise."

He managed the occasional weekend away—Morea owned a six-seat Cessna 210, and would fly Angela and their two kids down to a Florida beach for some desperately needed time with the family. But always, he had to dash back in time to report to work on Monday. "I barely had time to get in the water," he told me, "and I had to get back out." With his work on the J2 wrapped up, he had a welcome break—and thought it would be fun to visit a world's fair site. "I'd heard about Expo 67 up in Montreal, and I said to my wife, 'Hey, why don't we fly up there and see what that's all about? We can find ourselves a hotel someplace and spend a few days up there.' She thought that was a great idea, and we took the kids. So we got up there, and I checked into the hotel. And there was a message for me from von Braun's office."

It was from Erich Neubert, who'd been with the boss since Peenemunde and was now the Marshall Center's associate deputy director for technical matters. When Morea called back, Neubert said the center had just been assigned a new task and that Dr. von Braun "had indicated that I might be a good candidate to head the thing up," Morea recalled. "I said, 'Oh? What are we talking about?' He said, 'It's a car that we want to provide the astronauts.'

"I said, 'Well, wait a minute. I've been involved in propulsion at the center for my entire career, just about—for ten years now. I know something about propulsion. I sure as heck don't know anything about a car, especially one that's going to drive on the moon.'" He begged off the job. Neubert said he'd relay his comments to the director.

He called back the next morning. "He tells me, 'Dr. von Braun says that for a guy who's run the F1 program, and who fixed the J2, building a car shouldn't be any trouble at all,'" Morea said. "'He thought this project would be a piece of cake for you.

"'Besides,' he said, 'you only have to worry about it for a few months.'"

Morea flew his family back to Huntsville. He had no time to waste.

31

NASA PROCUREMENTS FOLLOWED A SCRIPT. FIRST, A REQUEST FOR proposals had to be prepared and sent to prospective vendors—in this case, the aerospace companies that the Marshall Center judged worthy candidates. Interested companies were given a statement of work. When they responded with proposals, a Source Evaluation Board would weigh them and recommend a winner. NASA's administrator, Thomas Paine, would make the final call. At that point, the selected company and the government would sign a "letter contract" enabling them to get to work, even as they continued to dicker over the terms of a formal "definitized" agreement.

All of this typically took a good many months. This time the process had to be fast-tracked, despite the many players involved just within NASA. Astronauts and mission planners were keenly interested in the vehicle's attributes, so the Manned Spacecraft Center in Houston had a voice—and Houston was vocal in its skepticism about the whole idea, as the rover's added weight meant less fuel on the lunar module and a slimmer margin of error on landings. The Kennedy Space Center had a piece of the action because the finished rover would have to arrive at the launch complex according to the Cape's schedule, fit into the lunar module without drama, and sit in the stacked Saturn V for weeks before launch. Headquarters had opinions galore.

Morea's procurement plan, put together with all this help, illuminated just how breakneck the schedule would be. The Marshall Center would release its request for proposals on July 11 and receive them just six weeks later. The evaluation board would have nineteen days to work. Administrator Paine would select the winning bidder about three weeks after that, on October 2.

Because the project involved a new and untried piece of hardware, conceived and built at an unprecedented pace, Morea and his bosses agreed that NASA would offer a cost-plus-incentive-fee contract to the winning bidder. That company would be reimbursed for all of its actual expenses and earn a profit commensurate with how well it met the schedule, the cost targets, and—most importantly—how well the resulting rover performed. In theory, the arrangement gave the contractor strong incentive to work smart and fast.

True to schedule, the request for proposals was released on July 11. Twenty-nine companies received the invitation to bid, among them the usual suspects—Boeing, GM, Grumman, and Bendix—as well as others that Marshall thought up to the job, such as Bell Aerosystems and North American Rockwell, which were still tinkering with their lunar flyers; Chrysler, which built the Saturn I's and IB's booster stages; and Philco, Westinghouse, and Huntsville's own Brown Engineering.

The document laid out a daunting list of particulars for the finished machine. Building on the expectations voiced earlier by headquarters, it had to have four wheels of flexible design, each powered individually;

carry two astronauts; draw its power from batteries; and carry 100 pounds of scientific gear in addition to its passengers. Because each astronaut would weigh an estimated 370 pounds in his space suit, that added up to 840 pounds, more than twice the rover's prescribed weight. And that was before the crew piled as much as 70 pounds of lunar samples aboard. Just as a reference, cars and trucks on Earth are rarely called upon to carry more than *half* their weight.

The rover's speed had to be continuously variable, with a top end of ten miles per hour over smooth ground. It would also need some bighorn sheep in its genes: NASA required that it be able to ascend a twenty-five-degree slope fully loaded, climb over a foot-high (thirty-centimeter) obstacle from a standing start with one set of wheels already resting on the obstacle, and clear a crevasse more than twenty-seven inches (seventy centimeters) wide, starting with its wheels in the crack and again from a full stop. It also had to be stable when parked sideways—and fully loaded—on an incline of forty-five degrees.

It had to do all of this while enduring a seventy-eight-hour lunar stay, which was the maximum NASA foresaw for the late Apollo flights. During that period, the car had to be ready for four sorties of 18.6 miles (30 kilometers) each, for a total of nearly 75 miles. It had to be designed with redundancies such that if any component failed, it wouldn't kill the mission, and if two went bad, the astronauts' safety wouldn't be at risk. Finally, it had to be folded and stowed so that a single astronaut could deploy it with minimum effort.

NASA made an additional request. Its enormous Michoud Assembly Facility in New Orleans was underused, and it wanted to know how the winning bidder might put it to use. The agency had acquired the complex in 1961, after von Braun recognized that its canal access to the Mississippi River made it ideal for building the Apollo program's giant boosters. Ever wonder how the biggest pieces of the Apollo rockets reached the Cape? Answer: by barge, from Louisiana. The Saturn V's second stage, built in California, was delivered to

Florida via the Panama Canal. The third flew by cargo plane from the West Coast.

In the summer of 1969, with the budget trimmed and the demand for Saturn boosters at a pause, NASA was "looking for ways to keep the doors open," Morea said. Building the rover at Michoud wouldn't employ nearly the number of people the rockets did, but it would help. And the place offered a second plus: Marshall managed the facility, and it wasn't far from Huntsville. It would be easy to keep an eye on the project.

32

EVEN BEFORE IT HAD SEEN THE REQUEST FOR PROPOSALS, BOEING was gunning for the contract. The company's point man was Herman Newman, the chief of engineering design at Boeing's Huntsville offices. Nowadays, Huntsville has the feel of a second Silicon Valley; a swath of the city is the province of tree-shaded technology labs drawn there by the Marshall Center. In July 1969, however, Boeing and most other aerospace firms were all working in a former cotton mill a few miles from the space center: the Huntsville Industrial Center, or, as everyone called it, the "HIC Building," pronounced *Hick*. "Herman went out to Marshall one day," recalled Boeing engineer Al Haraway, "and he came back saying that NASA was going to put out bids on an LRV. Herm had a conversation there, and he came back and said, 'We've got to try this.'"

Henry Kudish, Boeing's chief engineer in Huntsville, and Newman's boss, marshaled his meager forces to put together a proposal. That they'd partner with General Motors in the venture was a foregone conclusion. They brought complementary strengths to the job, as they had demonstrated in their past NASA projects. "Each of these programs was concluded satisfactorily and demonstrates an ability for Boeing and GM to communicate and work effectively together," as Boeing told NASA.

GM's Santa Barbara lab had gone through yet another reorganization and name change. Now it was part of AC Electronics, the GM division headquartered in Milwaukee, and was called the AC Defense Research Laboratories, or AC/DRL. The shift distanced it from GM's corporate leaders in Detroit, which might have been a good thing. By now, those leaders "were not very interested in or supportive of the lunar rover space program," Frank Pavlics told me. "They didn't trust our small laboratory. They said, 'Well, if there's an accident and the thing turns over, everybody will know that's a GM product. It's a risky business.'"

Sam Romano was undeterred. "If there's going to be a vehicle on the moon, it's going to be a General Motors vehicle," he'd say later of his attitude. "And I'm going to make sure that happens."

Truth be told, neither Boeing nor GM could have competed for the contract without the other. Pavlics and his team had built cars and parts of cars ostensibly designed for space, but never anything that was ready to strap to a rocket. Boeing was among the world's great systems engineering outfits and knew well how to build a spacecraft—but it hadn't built a car. Together, however, they were formidable. Pavlics figured that the partnership, along with his demonstration of the model in von Braun's office, made them a shoo-in.

He was in for a rude surprise, for the competition was stiff. Bendix and Grumman had been tinkering with their own shrunk-down versions of their LSSMs, and each had built a working full-scale Earth prototype. On top of that, the Bendix design was well known to von Braun. Photographs abound of him at the controls of the company's mobility test article, essentially an economy-sized version of its proposed rover. And on a visit to Grumman's Long Island operation, von Braun had ventured to the four-acre moonscape near Calverton for a test-drive. He "had such a good time driving the moon car" into and around the craters "that we couldn't get him to stop," a Grumman engineer told the *New Yorker*. "He wound up on an airplane runway about a quarter of a mile off, and the men in the control tower had to stop air traffic so he wouldn't be hit by an incoming 707."

33

ON JULY 16, 1969, FIVE DAYS AFTER MOREA RELEASED THE REQUEST for proposals, Apollo 11 launched for the moon from an oceanfront pad at the Kennedy Space Center. When the lunar module *Eagle* touched down, and Neil Armstrong and Buzz Aldrin stepped from its relative safety into the Sea of Tranquility's "magnificent desolation," rover teams at companies scattered across the United States were racing to assemble proposals by the August deadline.

Apollo 11's brief lunar visit offered a strong argument for the LRV project. Moving in a space suit proved tiring and clumsy: the outer garment was twenty-one layers thick and resisted bending as a matter of course, but when pressurized, it blew into a stiff balloon that required real muscle to flex. "It stands away from you. It's almost like being in a really small room," recalled Craig Sumner, then a space-suit-wearing test subject at the Marshall Center. "To bend an arm, you're working against almost five pounds per square inch of pressure. That's probably the hardest work I've ever done. I've always stayed in shape, but that was *hard.*"

Add limited visibility—from inside his helmet, an astronaut could not see his own feet—and layer on the weirdness of one-sixth gravity, and a lunar hike was a lot more work than it looked, and not nearly as fun. The upshot was that Armstrong and Aldrin were unable to stray far. Toward the end of their roughly two and a half hours outside the lunar module, Armstrong jogged to the lip of a nearby crater to take pictures and collect a few last-minute samples. That unscripted foray took him to the farthest point either man ranged from the lander— about sixty-five yards.

But the mission also fed Houston's unease about burdening the lunar module with any extraneous weight, including a rover. As the *Eagle* descended to the lunar surface, its automatic pilot aimed for a landing spot that Armstrong deemed unsuitable. He took control of the craft, skimming it over a crater and beyond, until he found a place he liked. In the process, he drained the lander's tanks nearly dry. The *Eagle* landed with just seconds to spare.

The next-generation lunar module planned for the later Apollo missions would carry more fuel. Even so, the added weight of a rover would cut into the time it could spend hovering—the formula bandied about within NASA was that every ten pounds of cargo cost a second of hover. Officials in Houston, especially Jim McDivitt—a Gemini 4 and Apollo 9 astronaut who now led Houston's Lunar Landing Operations office—argued in the name of safety that hover time was too dear to trade away. The Apollo 11 experience seemed to back him up.

So, even as NASA ploughed ahead with the procurement, the rover remained a fragile addition to the Apollo program. This became all too evident five days after Apollo 11's splashdown, when Armstrong and Aldrin sat down with NASA officials for their technical debriefing. The two "were questioned about using a roving vehicle," read a summary prepared for von Braun. The pair acknowledged the need for a rover, because "they estimated they could cover only one-half mile on foot without excessive fatigue."

But they had doubts, too. "They sounded pessimistic about surface roughness and the varying dust levels," the summary reported. What NASA higher-ups took from the meeting was "the feeling that perhaps LRV needs to be a different kind of vehicle than the current four-wheel approach." Maybe even a lunar motorcycle.

The Marshall Center's rover team tried to calm the managerial doubt, promising to show the astronauts a film clip of the Grumman prototype storming over rough country to demonstrate "what in fact these concepts can negotiate with regards to craters, slopes, rocks, etc." If Armstrong and Aldrin continued to denigrate the rover concept, then "we may well have to do some 'soul searching,'" they wrote.

Part of the problem, spelled out in later meetings, was that the astronauts apparently misunderstood how the rover was to get from one point to another. They figured it should cut a more or less straight path across the lunar surface, grunting its way into and out of craters, bounding over large rocks, when, in fact, its driver was expected to exercise some common sense and steer around them. The vehicle they had in mind would need wheels twenty feet in diameter, they said; short of that, they couldn't imagine it would bring much to a mission, because it would be no faster than an astronaut on foot.

That was flat-out wrong, as later experience would show. But no one, at the time, knew the lunar surface better than the two guys who had actually walked on the moon, and many at NASA took their comments as gospel. Morea opted to keep the bidding process on schedule and to stick with the requirements for the rover as they were written. But he also arranged for Marshall's soil experts to reassess their models for the moon's surface and to take the Bendix and Grumman prototypes out to Cinder Lake for another, harder series of tests.

In late September 1969 a contingent of high-ranking NASA officials assembled outside Flagstaff to watch the prototypes. Present were Manned Spacecraft Center director Robert Gilruth, Jim McDivitt—newly promoted to manager of the Apollo Spacecraft Program—and George Mueller, chief of the Office of Manned Space Flight at headquarters. "The mobility capability demonstrated by these vehicles greatly impressed the group," Morea, who also attended, reported to von Braun. "The capability to negotiate deep craters, slopes, and rocks was well demonstrated by both vehicles, and, generally, the entire group was favorably impressed." Gilruth, McDivitt, and Mueller drove the machines themselves.

That ended talk about killing the rover program. For the time being.

34

ON AUGUST 22, 1969, THE DEADLINE FOR ROVER PROPOSALS, packets from four companies arrived at the Marshall Center. To treat

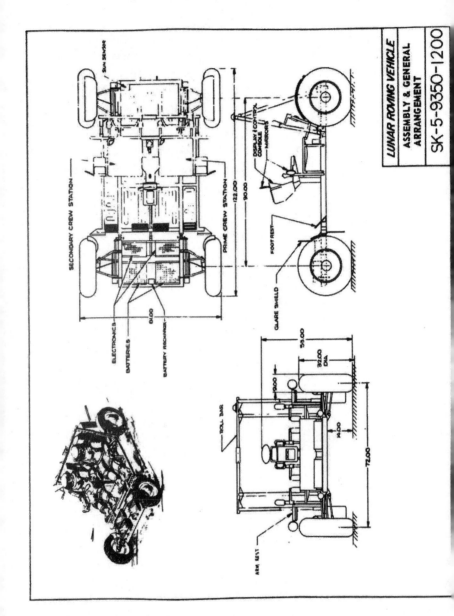

The proposed Boeing/GM rover. (NATIONAL ARCHIVES)

all of them in equal detail here would require a book by itself, so instead, we'll focus on just one: the pitch from Boeing and GM, which in its level of detail and thoughtfulness was representative of all.

At nearly two inches thick, the proposal contained a ridiculous amount of information for what might seem a glorified go-cart. Boeing's proposed rover was an open, four-by-four scrambler fashioned from welded rectangular tubes of aluminum alloy. Tiny by any standard, it had a wheelbase of seven feet, six inches, and an overall length of ten feet, two inches. For a sense of just how small that is, consider that a modern Mazda MX-5 Miata, among the smallest cars on the road, would outstretch it by more than two and a half feet.

We'll break the machine down into pieces.

The Frame

The chassis's tubing and rounded corners made for "a clean geometry that will prevent hang-ups during deployment and operation," the proposal explained. They also served "as conduits to protect wiring against damage," and their lack of sharp edges would "minimize the possibility of tearing the astronauts' clothing." The chosen alloy, a copper-aluminum blend dubbed 2219, was strong and easy to weld.

The frame took the form of three rectangles. The central piece, nearly square at fifty-three inches long by about fifty-six wide, supported the two-seat crew station. The smaller forward and aft sections were narrower by about a foot and a half to accommodate the suspension and wheels. The deck covering the main section was a panel of 2219 alloy just one-fiftieth of an inch thick—a half millimeter, or about as meaty as a particularly slim wood veneer—but stamped with ribs to boost its load-carrying strength and the frame's overall stiffness. It had the muscle of a flat panel twice as thick. Still, if a space-suited astronaut were to stand on the deck on Earth, it would buckle under his weight. Like all of the Boeing/GM concept, it was designed solely for the moon's one-sixth gravity.

The Crew Station

The astronauts sat shoulder to shoulder in seats that folded flat when the rover was attached to the lunar module. The seats themselves looked a lot like beach chairs—they were framed in tubular aluminum, with backs of crisscrossed webbing made of Teflon-coated, fire-resistant nylon. They looked uncomfortable, and so they would have been on Earth: they were exceptionally deep to accommodate the astronauts' hulking personal life support system backpacks, which contained their radio hardware and circulated air and cooling water into their suits while they were outside the lander. A gap between the seat cushions and backs was designed to accommodate the packs, helping to stabilize the astronauts while the rover was under way. The seats were equipped with nylon safety belts fastened with Velcro. Adjustable raised footrests enabled the astronauts to brace themselves while bouncing over rough lurrain.

Rising from the floor between the footrests was a pylon topped with a blocky control and display console. Its gauges and dials could be read by either astronaut. Extending aft from the console, just below the instrument panel, was a small shelf on which was mounted a pistol-grip joystick and an armrest.

As proposed, the joystick was a close match for the stick used to control the lunar module on its descent. Positioned so that either astronaut could use it, the stick worked a lot like an airplane's. Push it forward, and the rover motored straight ahead. Tilt it to the left or right to turn. Pull it back to brake. Pull it all the way back, and a parking brake engaged. Flip a switch on the handle, then pull back, and the rover would shift into reverse; an oval rearview mirror jutting from the console's top would show the way.

Two tubular aluminum handholds rose like bull's horns from the console's sides. Additional handholds were built into the frame outboard of the seats. And if the astronauts' haul of rock samples overflowed the seven square feet of storage in the rover's tail, they could remove the right-hand seat for an additional ten square feet of cargo space.

Navigation and Instrument Panel

Electronics piled inside the console served as the rover's brains, including the complex navigation system Boeing envisioned. The company promised it would enable astronauts to pick a destination and drive there or to return to a previously visited spot. It also kept track of where they were in relation to the lunar module, so, in case of emergency, they could return to base by the most direct route. Once the system was fired up and calibrated at the start of each journey, the console would tell the astronauts everything they needed to know— "The bearing to the LM, distance to LM, vehicle heading, distance traveled, vehicle speed, vehicle coordinate positions, and pitch and roll vehicle attitude." And it would do this accurately: if they set off for base camp from 3.1 miles (5 kilometers) away, it would get them there with an error of no more than three degrees of bearing and 10 percent of distance—or well within sight of their destination. The instrument display also included gauges monitoring the temperatures of the motors and batteries, as well as the life left in the latter.

The Drivetrain

Boeing proposed a brush-type, series-wound DC electric motor to power each wheel. The 28-volt, variable-speed devices would produce one-quarter horsepower each, so that together they boasted the motive brawn of a weed whacker. This dribble of power didn't promise much speed—the rover was designed to top out at about ten miles per hour on flat ground—but translated, Boeing promised, into enough torque to meet NASA's requirements for climbing and obstacle clearing. More importantly, the motors had sufficient grit that, if two of them failed, the surviving pair could carry the astronauts to safety. In such a happenstance, the crew would manually disconnect the failed drives, letting those assemblies freewheel.

The most novel aspect of the drivetrain was its harmonic drive, the elegant gear reducer that had infatuated Pavlics for years. Its circular spline had 160 teeth, and its flexspline, 158—so with each complete revolution of the motor shaft, the circular spline (and the wheel to which it attached) advanced just two teeth, or one-eightieth of a turn. Because this eighty-to-one reduction was constant, the astronauts need not shift gears. And as it had on earlier GM designs, it would work this magic in an unbelievably snug, lightweight package: each harmonic drive weighed just a pound and a half, and fit inside a soup can.

The brushes in the motors were prone to wear, and standard lubricants would boil away in the vacuum of space, so the drivetrain was sealed in a single casing that was greased with silicone oil and pumped full of nitrogen. A solid-state drive control system would convert movements of the joystick to the motors and steering.

Wheels

The wheels were refinements of GM's earlier designs. Thirty-two inches in diameter and nine inches wide, they were as big as they could be and still fit in the lunar module. The tires were ringers for the pneumatic rubber models found on any terrestrial car, except that they were made of wire mesh—eight hundred strands of zinc-coated stainless steel piano wire, crimped at intervals of three-sixteenths of an inch, then hand woven. Each tire was mounted to a sixteen-inch spun-aluminum hub.

"The wire frame wheel can develop maximum traction without grousers," the proposal advised, "because the open mesh acts as grousers and causes the soil to shear. However, tread strips are applied to the wheel to prevent abrasion of the spring wires and to provide larger contact area for better flotation." As proposed, the tread would be made of aluminum alloy.

Inside the tire was the bump stop devised by John Calandro. The tire would deflect three and one-quarter inches before it was called into play. The proposal testified that the combination stood up to a beating: in tests, the GM wheel had survived an obstacle course while in a vacuum for ninety-four thousand revolutions, which was 40 percent more use and abuse than a lunar mission was expected to throw its way.

Perhaps the most amazing thing about the mesh wheel was that each, including the aluminum rim, weighed just twelve pounds and created a ground pressure of less than ten ounces per square inch. Boeing noted that it had tested this wheel, and only this wheel, for its rover. If it got the contract, it would build a closed version, covered with fabric, for comparison testing, and choose one or the other for the final design.

Brakes

Nothing fancy here. The design called for mechanical drum brakes, with two shoes per drum. In theory, the astronauts would not rely on them much. The rover did not coast; if the driver eased off the joystick, the motors' output would drop, slowing the vehicle. This "dynamic braking" would suffice in most driving situations.

Suspension

The four-wheel independent suspension had an impressive ten inches of vertical travel, which promised to smooth out many of the bumps it encountered. The suspensions, identical front and back, were of conventional double-wishbone design, with torsion bars in place of coil springs—a feature found on many tanks, trucks, and SUVs. Here they'd save space and weight while providing a softer ride. The suspensions were cushioned by hydraulic shock absorbers, which were filled with silicone oil for durability in the moon's temperature extremes.

Steering

As described in the proposal, the Boeing rover would have electronic power steering in its front wheels. It relied on an Ackerman linkage, which meant that its wheels were not parallel while turning. When steering to the right, for instance, the car's right wheel would cut a slightly tighter angle than the left, so that they traveled parallel arcs of the same center point. That kept both wheels on the ground, rather than skittering sideways. The car in your driveway has the same setup.

The steering rack was powered by a small electric motor. "The selection of power steering was made only after a careful investigation of manual steering," the proposal said. "Although manual steering would be simpler and cheaper, it would impose undue strain on the astronauts." For one thing, it would require more physical effort. For another, it would mandate a steering wheel, which would be tough to handle in a space suit, require separate throttle and brake controls, and couldn't be centered between the riders, eliminating the ability of the passenger to take control in an emergency.

One steering characteristic was sure to get low marks from NASA: Boeing's rover would have a minimum turning radius of twenty-one feet, more than twice the vehicle's length—a clear violation of the agency's requirements. Boeing acknowledged that it could have added steering to the rear wheels, thereby tightening turns, but decided to go with a single system "on the basis of simplicity, power consumption, and weight."

Power Supply

Boeing proposed fitting its rover with two rechargeable silver-zinc batteries weighing thirty-six pounds each and connected in parallel, so that the astronauts could draw power from one or both at once. Rated at 1,800 watt-hours, they would have more than enough juice to power the navigation, drivetrain, steering, and control and display subsystems, and to complete sorties with half their charge remaining. As proposed, they would be recharged between trips with a solar panel.

Thermal Controls

The least sexy aspect of the rover's design was also one of the most important, for the lunar environment presented temperature extremes that would attack virtually every piece of the vehicle, from welds, to its delicate electronics, to its high-speed electric motors. Without armor against heat and cold, the machine wouldn't last through a single sortie, let alone a mission.

Again invoking simplicity, Boeing's design used a passive defense. The drive motor casings would be fitted with horizontal fins "to aid in the radiation of heat from the motor and harmonic drive," as would the shock absorbers. The electrical components and console, which promised to generate their own heat, would be heavily insulated and fitted with mirror-backed quartz glass radiators. These would have covers to protect them and prevent overcooling whenever the rover was idle. The radiators required the astronauts' careful attention while the rover was under way: "Dust collection on the radiator may seriously degrade radiator performance," Boeing wrote. "Therefore, the baselines include weight allowance for a device to remove the lunar dust from the radiator surfaces." A brush, in other words.

The most fragile components, temperature-wise, were the batteries, which Boeing said couldn't be allowed to top 110 degrees. They, too, were fitted with radiators and covers. "During a typical sortie, the thermal mass of the batteries is sufficient to prevent overheating, even with the radiator covered," the proposal read. "However, repeated charging and sortie operation would continue to heat the battery and eventually require a long downtime for battery cooling." Though it was not mentioned explicitly in the proposal, the battery covers would be automated: after a drive, the astronauts were to leave the covers open; when the batteries had cooled, the covers closed automatically.

The rover's body needed protection, too, for just a few minutes' exposure to the lunar sunshine could make it too hot to touch, even through an astronaut's glove. Boeing proposed painting the entire vehicle with white silicone paint "to maintain temperatures below 250 degrees." Which still sounds pretty hot.

Stowage and Deployment Gear

The gear to carry and deploy the rover was "basically two separate systems," Boeing said. The tie-down subsystem lashed the machine to quadrant 1 at three attachment points and held it fast during launch, the journey through space, and landing. The unloading subsystem released the balled-up rover. To get things going, an astronaut would pull a release cable that separated it from the tie-down structure. He would then pull a series of lanyards, causing the vehicle to snap open like a switchblade and drop to the surface.

"The design is a reliable mechanical system, operates semi-automatically, and does not interfere with LM functions," Boeing claimed. "Actuation does not tax the pull strength of the astronaut, and safety is assured by locating the actuation station on the LM ladder." The unloading operation, the company said, could be "accomplished in less than ten seconds."

Roll Bar

Boeing's proposed rover had a roll bar the astronauts would attach to the vehicle once it was on the ground. It was made of aluminum tubing salvaged from the deployment structure on the lunar module. Thus, the company said, "additional astronaut protection is obtained without additional weight."

Weight

Boeing claimed its machine would weigh 392 pounds on Earth, including the gear that attached it to the lunar module and unloaded it on the moon. The company could go even lighter with bumps in price and hassle: the chassis could be fashioned from titanium rather than aluminum, though the substitute would be touchier to work with. Myriad other savings could be found by machining structural pieces and such. "These factors," the proposal said, "lend confidence to delivery of LRV flight hardware within the 400 pound (empty) mass constraint."

Those were the headlines. Boeing's proposal dissected each element in much greater detail, achieving such levels of granularity and technical complexity that experienced engineers had their hands full. The company planned to manage the project from its offices in Huntsville and assemble the rover at Michoud, where it already had "a large, active, and trained work force" devoted to building the Saturn V's first stage. GM would build the mobility subsystem—pretty much every piece that moved—in Santa Barbara. The thermal and vacuum testing of individual parts and complete vehicles would happen at Boeing's Kent Space Center outside Seattle.

Boeing expressed great confidence in its partnership with GM and boasted that the companies had already answered the tough questions about how a rover should work. "Major portions . . . have been fabricated and extensively tested," the document declared, adding that such know-how afforded "a high level of confidence for the successful achievement of a qualified final design." Many elements were "essentially complete." As for those pieces that weren't yet in the bag, Boeing planned to use "off-the-shelf hardware items, proven components, materials, and manufacturing processes," so that "development [is] kept to an absolute minimum."

The bottom line, Boeing said, was that its team could build the thing on time.

35

ONE MIGHT ASSUME THE OTHER BIDS ON THE JOB WOULD BE VARIA-tions on the same theme. But such wasn't the case—only Boeing used an Earth car's layout. We'll tackle the remaining three from most orthodox to least.

Bendix

If you'd been hanging around a gravel pit west of Ann Arbor in the summer of 1969, you might have seen for yourself an Earth-ready version of the rover that Bendix proposed to NASA. The design was familiar to the space agency, in that it shrank and streamlined the moon vehicle concept the company had been tinkering with since its mobility test article. All of the idea's iterations relied on Bendix's signature elastic hoop-spring wheel—the mobility test article's at eighty inches in diameter, the Specified LSSM's at forty-three, and now, for the rover, thirty-two—with its circular springs and rolling surface of titanium sheet metal. Each had a short chassis with its wheels on stout, shock-dampened arms that extended fore and aft. As on Bendix's LSSM, the arms folded backward toward the vehicle's middle, creating a snug bundle for stowage.

By now, Bendix was an old hand at building Earth-gravity representations of its ideas and testing them out in the field before offering proposals to NASA. A gravel company had permitted the firm to bull-doze its pit into a moonscape of craters and mounds, and, over several visits that summer, Bendix test-drove its prototype there, focusing on a different facet of its operation in each session. The trips eventually at-

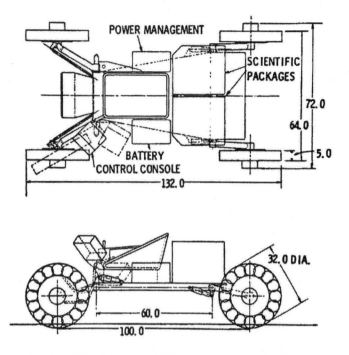

POWER MANAGEMENT

SCIENTIFIC
PACKAGES

72.0

64.0

5.0

BATTERY
CONTROL CONSOLE

132.0

32.0 DIA.

60.0

100.0

An early drawing of Bendix's concept, this one seating a single astronaut but in many respects a ringer for the company's LRV proposal. (NATIONAL ARCHIVES)

tracted a reporter from the *Detroit Free Press,* who wrote about the rover in a story headlined "Weird Car Begins Race to Moon in a Quarry."

Weird or no, this was a viable candidate for Apollo. It weighed 394 pounds, Bendix claimed, and deployed from the lander with a hand crank—simpler, and presumably more reliable, than the semiautomatic system proposed by Boeing. It stretched eleven feet long, on a wheelbase of eight feet, four inches—both about ten inches longer than Boeing's. It didn't have a roll bar, but then, Bendix claimed it could park sideways on a slope of more than fifty degrees without tipping.

The rover's front and back wheels pivoted in opposite directions at commands from its joystick controller, enabling it to turn in its own length. That joystick was centered in the crew station, which featured side-by-side seats. The astronauts climbed aboard by stepping backward into the front of the machine. The crew station also contained room for a spare backpack.

Bendix's navigation system was complex, perhaps more so than the mission required, employing multiple directional gyroscopes. It would be initialized—that is, the rover's position fixed at the start of each sortie—remotely from Mission Control.

Its electric motors were stepped down with a nutating transmission, which used elements that intentionally wobbled to engage only some of the time—achieving much the same effect as the harmonic drive's wave generator, though with more moving parts. Like its predecessors, the Bendix rover was no racehorse. It sauntered from zero to ten miles per hour in fourteen seconds.

Chrysler

The proposal from Chrysler Corporation's Space Division, out of New Orleans, was outclassed from the start. Its astronauts sat back-to-back, so that only the driver could see where they were headed—a detail that was bound to disquiet NASA. Not only that, but driving was a two-handed operation: Chrysler proposed separate controls for speed and steering, the latter a T-topped handle between the driver's knees. Oh, and at deployment, the astronauts would have to attach the wheels. Considering that the rover project was aimed partly at preserving the astronauts' time and energy, that might have been a deal killer right there.

The 398-pound rover was skeletal, its frame a single sturdy aluminum spine linking the axles, augmented by two girders extending from the rear wheels in a wishbone to meet beneath the front seat; if you stood it on its tail, the arrangement looked like the inside of a peace sign. Together the three girders provided the floorless rover with support for both seats, as well as the machine's electronics and batteries. The front third of the central backbone telescoped into the rear, shortening the car for storage in the lunar module—a nifty piece

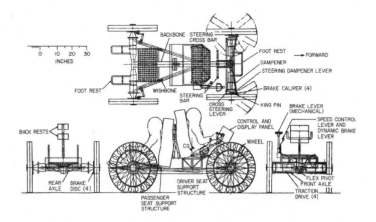

Chrysler's tiny, tandem-seat rover concept. (© SAE INTERNATIONAL)

of engineering all but negated by that detail about the wheels, which would be stored separately in quadrant 1.

The Chrysler machine was far narrower than the competition, thanks to its tandem seating. The approach left little, if any, room for tools and geologic samples—for anything, in fact, but the vehicle's own running gear and the astronauts, who boarded the rover from opposite ends, performing an about-face to take their seats. The passenger sat several inches lower than the driver, with his feat in stirrups that trailed the machine's rear. Chrysler's wheels mimicked those developed by Bendix. Like Boeing, it relied on harmonic drives to step down its electric motors and used Ackerman steering only in the front wheels, so that it required two and a half times its own length to turn. Its rechargeable silver-zinc batteries would be refreshed, like the proposed Boeing rover, with a solar array.

Alone among the entries, the Chrysler boasted disc brakes—cable activated, as on modern high-end bicycles. Its temperature-management system drew excess heat from the components and transferred it to a reservoir of water. When the water boiled, the rover vented the resulting steam. Only in one respect did the concept offer a clear advantage over the others: its navigation system was the least complex of the bunch.

Grumman

The most interesting proposal of all might have been Grumman's. It was faithful to the concepts the company had developed in its studies for the LSSM, and the crazily nimble prototype that von Braun test-drove on Long Island. Once again it relied on pliable fiberglass-resin wheels shaped like bowls, with rectangular titanium cleats along their outer edge. Under the rover's weight, the wheels flattened on the bottom, increasing their ground contact area and thus their traction. Their flexibility also absorbed jolts. A reporter for the *New Yorker* described seeing them in action on the prototype: "The moon car hit a biggish rock, which punched the wheels completely out of shape. For a second, the car looked as if it were riding on squashy melon rinds."

The Grumman rover was the widest of the entries, at nine and a half feet; the Boeing design was nearly three feet narrower. That span virtually immunized Ed Markow's machine from flipping over, despite its nineteen-and-a-half-inch ground clearance, and lent it the air of a crouching spider. Its chassis was its most daring element: it was articulated, with two pods linked by a flexible spine. The forward pod contained the crew's seats, which were side by side and entered from the front. Between them were controls for either occupant. The rear pod carried the batteries, tools, heat radiators, and payload.

When its joystick was pushed into a turn, the wheels didn't change orientation—the rover bent in the middle. The feature enabled Grumman to eliminate steering racks and complicated wheel attachment rigs, dramatically reducing the vehicle's complexity. Another plus: the arrangement ensured that the back wheels would follow the tracks created by the front, reducing power consumption. And if the steering failed, the controls could be disabled, the link between front and back locked stiff, and the rover's direction changed by varying power to the wheels on one side or the other. It was a design after Greg Bekker's own heart.

Grumman's rover had rechargeable batteries, juiced between sorties

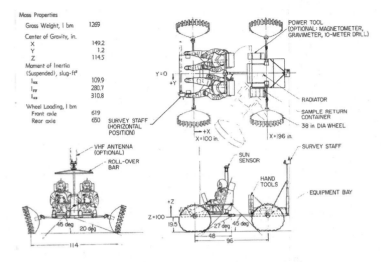

Grumman's strange and wonderful proposal. (© SAE INTERNATIONAL)

by tapping into the lunar module's reserves. Those batteries would run a complex navigation system as well as the four independent electric drive motors, which were stepped down with traditional planetary-gear transmissions. Heat generated by the motors and electronics was transferred to a sink of wax under the radiator at the rover's tail. The melting wax alone, Grumman claimed, would absorb 75 percent of excess heat during a sortie. The radiator would do the rest.

About the only downside to this remarkable idea was its weight: Grumman estimated it at 439 pounds. To openly violate one of NASA's hard-and-fast requirements was a dangerous move. Perhaps Grumman was so supremely confident in its craft and in its ability to put the competition to shame that it reckoned NASA would forgive the breach.

Then again, maybe it was just the most realistic bid of the four.

36

THE SOURCE EVALUATION BOARD INCLUDED NINE NASA EMPLOY-ees from Huntsville, one from the Kennedy Space Center, and one from headquarters. Its criteria for the competition gave the bidders marks for austerity, speed, and simplicity; concept feasibility; ingenuity ("Freshness of approach and objectivity should be emphasized"); and experience ("The bidder should have been involved in some effort associated with lunar roving vehicles"). Familiarity with the Marshall Center was important, too.

The board questioned the bidders both face-to-face and in writing. This back-and-forth sparked refinements. Boeing/GM, for instance, acknowledged that its rover's turning radius was too wide and agreed to add rear-wheel steering. In some cases, however, bidders didn't much help themselves. Representatives from Chrysler, which had nearly a half century's experience building cars, mentioned that its Detroit brain trust wouldn't have much of a hand in the project: the bid was from the company's Space Division. There were no flies on that outfit, but the board was mystified: why wouldn't Chrysler tap its deep well of expertise?

In the end, the panel ranked the Bendix proposal first. "It had an experienced team available, strong technical capability and organization," as Administrator Thomas Paine wrote later. "The mobility design was excellent, with double Ackerman steering providing the best maneuverability." Bendix's proposal was the most expensive, however, and not by a little.

The board assigned second place to Boeing/GM for its "several good features," along with its economy—it was the second cheapest of the lot. The chief qualm was that Boeing's work would be scattered hither and yon; its offer to base assembly at Michoud had, ironically, counted against it, because it also planned work in Huntsville, Santa Barbara, and the Seattle area.

Grumman placed third. "The proposal provided for good maneuverability and crew station design," Paine wrote, but the fiberglass wheels made everyone nervous, and Grumman had flouted the rules—its machine was overweight, and its electrical system exceeded the assigned limits.

The Chrysler entry was the lowest in price, but the board "had a low confidence in the capability of Chrysler to deliver a good vehicle on schedule within those proposed costs," Paine wrote. Besides, "Chrysler's proposed design was not acceptable as presented," especially the back-to-back seating arrangement. A host of smaller flaws were not easily fixed.

While the board reviewed the candidates, Sonny Morea performed his own assessment, developing a chart on which he assigned a number grade to each in several categories. Some were clearly subjective, such as "general attitude." Some were a wash—all four companies would have trouble meeting the weight requirement, he reckoned. But on most, he had clear favorites, backed by evidence. On vehicle stability, he rated the Grumman design tops, and Bendix a close runner-up, adding: "This will probably be a very important parameter if what the astronauts discussed in the Apollo debriefing is correct." On wheels, he gave the edge to Boeing/GM, "based on the fact that we know a great deal about this design." On depth of design, Bendix and Grumman led, as they both had working Earth-gravity prototypes. On safety, ease of getting into and out of the seats, deployment methods, and location of driving controls, Grumman took first place ahead of Boeing/GM.

Grumman prevailed in one category that Morea thought particularly important: technical interfaces, thanks to the company's authorship of the lunar module. It knew quadrant 1 better than anyone, and if it did *not* get the job, NASA would find itself with a failed bidder controlling much of the project's fate. In sum, Grumman was Morea's overall choice. Bendix came in second, Boeing/GM third. Any of the three could succeed, he wrote.

On September 29 the board presented its findings to Paine and two of his top lieutenants, Willis Shapley and Paul Dembling. Afterward, the three met with NASA executives who had a piece of the project, including Morea. "I kind of pushed it in the direction of Grumman," Morea told me. "I said that their proposal was probably more suitable because they'd been playing around with this, and they know what it'll take to incorporate it into the LM, to be able to take it to the moon."

He was overruled. Paine went with the board's top two finishers. "Mr. Shapley, Mr. Dembling, and I agreed that final negotiations with both Bendix and Boeing should be conducted," the administrator wrote. "It was our judgment that further negotiations with Grumman and Chrysler would serve no useful purpose and would not be productive."

At Grumman, Ed Markow took the news hard. "It was a bad day," his son, Jim, told me. "He was pretty hard to console on that, because they had put so much into it. They felt they had the best product, for sure."

Tight as the schedule was, the Marshall Center had only two and half weeks to tear through its technical discussions and contract negotiations with Bendix and Boeing, exploring everything from costs, to incentives, to design refinements. The outcome would be a letter contract with each company, from which Paine would choose the winner.

NASA spelled out its technical issues with the Boeing/GM proposal in early October. It complained that forward visibility from the seats wouldn't be great. Boeing responded by lowering the control console several inches and tilting it so that it blocked less of the view out front. The agency suggested that getting on and off the seats might be difficult in the astronauts' bulky pressure suits. Boeing replied that it would put the seats on swivels and move the entire crew station forward by 5.7 inches to put some space between the seats and the rear wheels, which looked as if they might get in the way. The government worried that Boeing proposed to spread the work among too many locations. Boeing scrapped its plan to use Michoud.

Boeing also agreed to trade its solar charging system for simpler nonrechargeable batteries, and to provide alternatives for two design elements that Pavlics and Bekker were particularly fond of: If the wire-mesh wheels didn't pass NASA's tests, the company would offer a different approach. And if the harmonic drive didn't work as advertised, it would use the Bendix transmission, which some in the Marshall labs privately preferred.

In short, Boeing addressed all of NASA's reservations with little or no fuss. That ushered in two days of formal contract talks, during which the company embraced the most complex piece of the agreement: the cost-plus-incentive plan. The arrangement that Morea sketched out gave Boeing a fee ranging from 1 percent to 14 percent of the target cost, or from as little as $173,000 to as much as $2.42 million. To reach that top figure, absolutely everything would have to go right. Better than right, actually: the finished rovers each would have to come in on time, underweight by twenty pounds, under budget by 20 percent, and perform flawlessly on the moon.

A more achievable fee fell in the middle of the range: 7 percent of the target cost, or $1.21 million, which required four successful missions but assumed that the weight and budget bonuses were beyond reach. That was decent money in 1969, puny as it seems now—close to $8.5 million of today's dollars. Perhaps more importantly, a good outcome offered bragging rights and promised more NASA work in the future. It seemed the proverbial win-win: Boeing's profits depended entirely on its own performance, and NASA could rest easy that by sharing the program's risks and rewards, the company was motivated to protect the government's interests as much as its own.

The discussions with Bendix didn't go as smoothly. The company came into the talks more experienced and invested than any other organization, private or otherwise, in building rover prototypes. But its negotiators haggled over the statement of work word by word and took regular time out to complain about its past treatment at NASA's hands. Bendix was also strangely resistant to at least one of the eight design changes NASA requested. When the agency asked that it include three different battery options, Bendix said it would—but did not. This happened twice. "It was only after a great deal of verbal exchange that Bendix finally responded to NASA's wishes," agency negotiator Kenneth M. "Mike" Grant reported, adding that it required "extraordinary effort on NASA's part to refrain from *telling* the contractor how to design *his* vehicle." All that aside, when the talks closed, Grant judged that NASA and Bendix had hammered out a letter contract that was "satisfactory in every respect." The company's contract was the equal or better of Boeing's.

Except, in retrospect, for one thing: in the wake of Apollo 11's success, and with the approaching Apollo 12 mission inspiring far less excitement, NASA's lunar program was suddenly yesterday's news. This, at a moment when it competed for federal money with critical social needs in America's cities and a lingering, costly war in Vietnam. NASA headquarters favored spending as little money as possible.

Both bidders were promising they could build the cars for a fraction of the Marshall Center's estimate. Nevertheless, the Bendix rovers would drain more from the treasury than Boeing's. The two bids were more than $5 million apart—$22.96 million versus $17.3 million. And with the expected fees, the gap widened by another million. With a 7 percent incentive payout, the Boeing contract would total $18.51 million. Bendix's would run $24.57 million.

On October 27 Administrator Paine, accompanied by Willis Shapley and Associate Administrator Homer Newell, met again with the Source Evaluation Board to hear its report on the negotiations. Both companies had responded well to the board's suggestions, they were told. "Dr. Newell, Mr. Shapley, and I questioned the Board," Paine recalled, "to ascertain whether or not it had any reservations concerning the ability of either company to perform the procurement." No, they were told: "While the Board had substantially more confidence in Bendix than in Boeing, it reported that Boeing had competent people working on the proposal; that General Motors, a major subcontractor, was knowledgeable and competent for this type of work; and Boeing had accepted the schedule."

When the session ended, Paine met again with Morea and other NASA executives. With his first pick for the job eliminated, Morea leaned toward Bendix. But when Paine asked him whether he thought Boeing could do the job, "I was hesitant to say that Boeing couldn't," he said. "They had certainly proven themselves in so many fields, and they had a tremendous engineering capability." His one nagging reservation was Boeing's cost estimate: he doubted the company could do the job for so low a number. "I made those comments directly to the administrator," he told me. "They said, 'Well, they're the low bidder on it. Let's go with that.'"

The news came as a shock to Bendix, which "was pretty sure they were going to get this LRV program," contracts negotiator Mike Grant recalled. "Their fellow in Huntsville, he came into my office after Boeing got the contract, and he actually cried, standing in front of me. He just stood there, crying. It was upsetting."

In Santa Barbara, Frank Pavlics and his GM colleagues celebrated. "Of course, we were excited," he said. "We'd been peddling this rover idea, and we'd finally succeeded. Now we'd get to build it.

"But then we realized: now we *have* to build it. Now we had left eighteen months to design, test, and build the first."

PART FIVE

A
PAINFULLY
TRYING TASK

37

GAZE UPON THE ROVER DISPLAYED AT THE U.S. SPACE AND ROCKET Center, spartan as it is, and it's easy to assume, as Sam Romano and Frank Pavlics did in late October 1969, that building it would be a simple matter. They were pressed for time, certainly, but they had sorted out most pieces of the machine over the preceding decade—and they would be using them on a vehicle far less complex than the six-wheeled jobs they had favored to date. The few elements they weren't building—namely, the frame and navigation system—were in Boeing's experienced, skillful hands.

NASA went into the job with the same level of confidence. This was a small project, as Apollo went—technically straightforward, safe, and cheap. Boeing enjoyed a global reputation for engineering excellence. GM had deep experience in rover projects. What could go wrong?

Well, quite a bit, actually.

First, the deadline: it changed everything. In addition to four flight rovers, Boeing was on the hook for several exact or near copies for various space hardware tests. One gauged the rover's response to vibration and jolts during launch and lunar touchdown. Another finessed its interfaces with the lunar module, and a third checked the integration of the vehicle's various subsystems. The fourth and fifth were one-sixth-weight units used to develop the mechanism that unfolded the rover and lowered it to the lunar surface. A sixth, called a qualification unit, tested the entire machine's hardiness for space travel. Boeing was also expected to produce an Earth's gravity, or 1G, trainer for astronaut driver's ed, and a full-scale model of the rover's overall geometry to sort out the ergonomics of the crew station. Finally, it had to come up with support stands, the deployment and tie-down gear, and installation equipment, along with a raft of documentation on how everything worked.

So Boeing had a lot to do in very little time, including a great deal of testing. Every component had to undergo development testing to nail down the details of its design, after which it would endure qualification testing to ensure it could withstand the privations of travel to the moon—all while it was subjected to other tests intended to break it and thereby establish its limits. These trials were normally finished before assembly began on the flight-ready hardware. For that to happen this time would require careful choreography, not least because while much of the testing could be performed in Huntsville and Santa Barbara, some could only happen in Kent, Washington, where Boeing kept its thermal-vacuum chambers and other complex simulation gear; even without Michoud, the project's scattered geography loomed as a complication. Only a few pieces of the job would be designed, tested, and fabricated in a single location—most notably the 1G trainer, which GM would build in Santa Barbara.

The second thing that might go wrong: the budget. It offered little, if any, cushion. Unforeseen problems could scramble the company's figures; a minor technical issue might invite delays that could be countered only with overtime, and thus exceed the job's fine tolerances for things going just so.

Time and money—those were the two fixed elements of the task ahead. Using too much of either would send the thing off the rails. Every facet of the project required meticulous management. We won't have to wait long to see how that turned out.

38

FROM HIS FIRST DAY AS PROJECT MANAGER, SONNY MOREA REC-ognized NASA's absolutism about the rover's weight as one of his toughest challenges. Four hundred pounds was a confoundingly slight mass for a vehicle that had to take a beating while hauling around twice its weight, especially when its batteries alone might account for a significant share of the total. Equally concrete was the payload limit for tools, gear, and geologic samples, a hard-and-fast 170 pounds. The stringent weight limits were necessary to avoid a spiraling dilemma, for exceeding them would require more of the rover's motors, which would demand more electrical power, which would require beefier batteries, which would add still more pounds.

The first threat to the weight edict emerged even before Boeing won the contract. In October 1969 the Manned Spacecraft Center in Houston insisted that the lunar communications relay unit it was developing had to be included in the rover's payload. Mounted on the rover's nose, the LCRU was a box stuffed with electronics enabling the astronauts to beam TV signals back to Earth when they were parked. It also relayed their voice transmissions and their biometric readings whenever they drove beyond the range of the lunar module's radio relays. If the rover broke down, the crewmen could unhook the LCRU and carry it like a briefcase as they walked back to the lunar module; it would keep them in touch with Mission Control.

No one questioned its importance. The problem was that the rover's weight had been figured without it, because it wasn't technically part of the vehicle—the relay unit was one of several add-on components, along with the TV camera and antennas, that Houston would supply and that the astronauts would attach once the rover was separated from the lander. And it was a big add-on: the LCRU was expected to weigh forty to fifty pounds. Morea recognized that the shift was basically an accounting trick. By having the rover absorb the extra pounds, the Manned Spacecraft Center wouldn't have to find an equal number to trim elsewhere on the lunar module. They'd be Huntsville's problem, not Houston's. And they'd have to come out of the rover's "already tight 170 pounds" of payload, he warned his bosses.

At almost the same time, Houston passed along word that the

rover's occupants would be wearing modified space suits and back-packs, boosting the total weight of each astronaut from 370 pounds to 400. This increase, too, had to be offset. Morea wrote Ben Milwitzky in Washington: "I don't feel we can continue to absorb new requirements and still be expected to meet our weight, costs, and schedules."

While Morea obsessed over the rover's unwanted pounds, the Marshall Center's labs were subjecting the Boeing design to minute and thorough scrutiny, and quickly came up with suggestions to strengthen it. They started with the rover's drive motors.

Any electric motor is in essence a magnetic device. Typically, current flows into the motor's shell, creating a magnetic charge there. The shell wraps around a rotor that is electrified to carry a charge, too, and the attract/repel character of the resulting magnetic fields spins the rotor, along with the output shaft to which it is attached. Reversing the charges every half spin keeps the movement going.

The brush motors specified by GM had been around for decades. Within the Marshall Center, however, some engineers worried that they weren't up to the job; Walter Haeussermann, one of von Braun's German colleagues and long the head of the Astrionics Lab, was especially vocal in his doubts. At issue were the brushes, which delivered current to the spinning rotor. Typically made of carbon or graphite, these brushes were prone to wear on Earth, and in moon's harsh environment, there was no telling how long they'd last. Or wouldn't.

It was the sort of interdepartmental meddling that von Braun encouraged: the labs' deep expertise and willingness to raise questions across organizational lines were key to rooting out technical issues early. Marshall's culture demanded that its people speak up. If you made a mistake, you told on yourself. If you found a flaw, no matter whose fault, you flagged it to your colleagues. And if you differed with a technical decision, you said so. When working around spaceships that can easily blow to bits, you couldn't be bashful.

So, though Boeing and GM insisted that brush motors would

work just fine, Morea instructed them to line up a backup. In early December, GM contracted with General Electric to build brushless permanent-magnet motors, which in 1969 were not nearly so common but didn't need to be sealed against a vacuum, thus making them simpler to install, at least in theory. This added insurance, Morea figured, would boost the project's cost by $500,000.

Haeussermann had another beef, this one with the Boeing navigation system, which he believed was far too complex. What was needed, he argued, was a bare-bones approach that brought the astronauts to within sight of their target, no more. After all, they'd be negotiating open ground with long sight lines and with maps to back up their electronics. If they somehow got lost, they could always turn around and follow their tracks back to the lunar module.

On this, Morea concurred. What might work, he thought, was a simple directional gyro mated to an odometer—just enough smarts to record the LRV's direction and distance from a fixed starting point, get it where it needed to go, and calculate the quickest way back. Houston chimed in with its agreement: keep it basic. Boeing wasn't happy with the change, but Morea brought it around.

The result was beautifully uncomplicated. Because the moon lacks a magnetic field, the system couldn't rely on a conventional compass to determine direction. Instead, it used a known starting point for each sortie—invariably, a few yards from the lunar module—and directed the rover toward its various destinations by calculating their relation to that point. Its single directional gyro, which recorded where the rover was pointed, was combined with the distance the machine traveled, a measurement supplied by odometers in each wheel hub. Those two streams of data were fed into a signal processing unit, which recast them into useful information on the instrument panel: a large compass rose showing the rover's heading, and dials displaying total distance traveled and the distance and bearing to the lunar module. Because some tire slippage was likely on the moon's loose soil, the system would always go with the odometer signal from the wheel spinning third fastest of the four. The new nav system was a huge improvement: sleek, sturdy, and reliable.

Other input wasn't as helpful. Headquarters asked the project team to figure out a way for the astronauts to use the rover's TV camera while seated in the machine. It was an odd request, as the astronauts weren't expected to do any shooting with the camera; plans called for Mission Control to operate it remotely, to capture the crew collecting samples, conducting experiments, and exploring the ground around the machine on foot. To make the change, the camera and the LCRU would have to be redesigned, and to what end? It wasn't as if a crew could shoot footage while the rover was on the move, because TV signals had to be aimed precisely at the earth. On the rover, they'd be transmitted through a high-gain antenna, shaped like an inverted umbrella and mounted on a long stalk on the vehicle's front end. Before turning on the camera, the crew had to use a sighting device on the antenna to ensure that the signals reached their target. Under way, with the rover bouncing over rocks and craters, a connection would be virtually impossible to maintain.

"It just doesn't seem possible," the project's engineers wrote to Morea. "To add such a requirement now is to ask for a major redesign just to incorporate a 'desirable' feature." Marshall's leadership agreed, and headquarters backed off the request, but not before the exercise had devoured time—and Morea had none of that to spare.

The job's first few weeks were otherwise aimed at getting everyone ready for the sprint ahead. Boeing and GM built their teams. To lead the project, Boeing selected the man who'd put together the winning rover proposal. Henry Kudish had been a career army officer before joining the company in 1957, and had worked since on its missile programs and the Saturn V's first stage. A few months before, Boeing had named him its chief engineer in Huntsville. Morea found him agreeable and well organized.

Herman Newman, who'd caught wind of the rover project before it went public, would serve as Boeing's top engineer. Another army veteran, he had come up on the aircraft side of the company, working on the B-47 Stratojet bomber and the 707, 727, and other airliners before shifting to the Saturn program in 1962. He had worked in Huntsville since 1967.

The first formal review of the project took place in mid-December at Boeing's Huntsville offices. This two-day preliminary requirements review, an early status powwow aimed at identifying problems likely to require special effort to surmount, yielded a few small changes. During tests of the crew station, Boeing had found that an astronaut could strap himself into the rover without swivel seats, and the parties agreed to junk them. They cut the rearview mirror attached to the console, too, as unnecessary weight.

Meanwhile, NASA and Boeing prepared to hash out the more detailed, formal pact cementing their partnership. It couldn't come soon enough for the Marshall Center because, within days of winning the job, Boeing started hiking its projected cost figures. On November 12 the rover's cost jumped a quarter-million dollars, from $17.3 million to $17.56 million. Twelve days after that, it rose again. The ratcheting price seemed to confirm Morea's earlier unease about the winning bid, especially when, in December, Boeing hiked its price past $20 million.

NASA pushed back. After a long back-and-forth, both parties settled on a new, everything-included target of $18,673,000. The incentive arrangement underwent some minor tweaking—Boeing's likely fee was raised from 7 percent of target cost to 8.5, for instance—but the merciless bottom line remained in place. If the project busted its budget, or ran late, or the rovers failed to perform, Boeing's fee could dwindle to just 1 percent of the target cost. NASA negotiator Mike Grant called it "the most stringent discipline one could possibly expect for this type of program."

39

NO GOVERNMENT PROJECT HAPPENS IN A VACUUM, AND A PAIR OF grim distractions soon presented themselves. The first came four days into 1970, when NASA headquarters canceled Apollo 20. Feeling the effects of its shrinking budget, the agency's leadership saw no hope of reopening the Saturn V production line at Michoud. It possessed only those boosters already built, one for each of the remaining Apollo flights, and decided it needed one of them to launch Skylab, a space station scheduled to follow the moon shots. The rover's completion deadline did not change, however. It was now slated to fly first on Apollo 16; that mission was pushed back to July 1971, into the launch slot originally reserved for Apollo 17.

The decision didn't come as a complete surprise. NASA had seen its lunar budget trimmed repeatedly over the previous five years. Beyond that, Apollo 11's success had already accomplished America's stated goal in the space race: putting men on the moon and returning them home safely before the end of the decade, and doing so ahead of the Russians. George Low, NASA's deputy administrator, pointed out that further Apollo cuts would squander the immense investment already made in the program, as well as limit its scientific returns. But in a note to his staff, Morea confirmed the program was in trouble. "Indications are bad for FY-71," he wrote, referring to the federal budget that would come into play later in the year, "further jeopardizing Apollo 19." A few days later, the weekly notes to von Braun included an ominous entry from Roy Godfrey, manager of the Saturn Program at Marshall. "A budget cut of $31 million was identified in Saturn Apollo funding for FY70–71," he wrote. "Initial studies indicate that some program deletions are required."

The angst that attended the withering budget was secondary, however, to another blow. In late January 1970 news leaked to the press that von Braun was leaving Huntsville for a planning post at NASA headquarters. The center's charismatic director, on vacation in the Caribbean at the time, had wanted to break the news to his colleagues himself.

Instead, confusion and anguish gripped the Marshall Center for days. On his return, von Braun, wearing a beard he'd grown in the islands, confirmed that he'd begin his new job in Washington on March 1. "I am leaving Marshall with nostalgia," he told his colleagues. "I have my heart in Marshall. I love this place. I helped build it up. I feel I'm a part of it." Nevertheless, this move stood to benefit everyone, he told them: he'd be able to help chart NASA's course beyond Apollo, beyond the moon, and thereby build on everything they'd worked together to achieve. "I've spent ten years doing what was urgent," he said, "and regrettably, not doing what was essential." The new job would change that.

Von Braun's handpicked successor was Eberhard Rees, his chief deputy since Peenemunde. The two had made an effective team: Rees, sixty-two, lacked the boss's charm, imagination, and flair for stagecraft, but he was practical, particular, and a first-rate engineer. Von Braun had high-flying ideas; Rees saw them through. Von Braun looked to the future; Rees tended to details of the here and now. Von Braun inspired; Rees managed.

His appointment, and the continuity it promised, was comforting. Even so, with von Braun's departure, a profound grief settled over many in Huntsville. Fifty years later, Sonny Morea had an almost dreamy cast as he described his former boss. "He was a brilliant man," he told me, "but beyond that, he was a brilliant manager of people, with a real sensibility of their psychological needs. People were the most important component of von Braun's career.

"I'm just so fortunate to have been exposed to that. I think my whole concept of management and leadership would be different if I hadn't worked with a giant like that. I owe him my career. He selected me from a lot of candidates and put me in charge of the F1 engine, and he supported me all the way.

"And, of course, he put me in charge of the LRV."

With von Braun's transfer to NASA headquarters, we say goodbye to one of our story's main characters. He was the conceptual father of lunar mobility, dating back to the *Collier's* stories of 1952. He rode herd on the Marshall Center's many studies into how best to design a vehicle for the moon. Not least, he'd witnessed Frank Pavlics's demonstration of the model and had a presumed hand in NASA's revived interest in the rover.

Once in Washington, von Braun was never put to use as he expected, especially after Thomas Paine left NASA's leadership a few months later; he became so disillusioned that he left himself, in 1972, for a post at Fairchild Industries, an aerospace contractor in Washington's Maryland suburbs. Minus his vision, his consensus style of leadership, his force of personality, and his public standing—and notwithstanding his deeply troubling past and Eberhard Rees's own strengths—Marshall was never quite the same.

Morea did not have time to wallow, however. The news leak about von Braun's departure came a day before the rover's preliminary design review, an important milestone. The review represented a major winnowing of options. Here the space agency and Boeing would agree on the basic design approach for each component and subsystem—or, in a few cases, agree to pursue parallel development plans until the best option won out. They would decide on the documentation that would back up the work. They would identify all of the interfaces between parts built by different subcontractors and figure out how to ensure they worked in concert.

Of the "open action items" unresolved at the review's end, the most important was weight: Boeing's current projection for the finished rover was 436.5 pounds. Morea and his team gave the company just under a month to come up with a plan to shed the surplus. Another pressing concern was the potential accumulation of dust on the machine. The two Apollo landings to date had established that, while the moon's surface appeared light gray in photographs, the dust staining the astronauts' space suits was nearly black, and tenacious stuff that smeared like graphite—and a dark coating on the rover's surfaces would absorb heat, throwing its thermal systems out of whack. With that in mind, the reviewers decided the machine should have a defense against a potentially major source of moondust: much more extensive fenders than originally envisioned, wrapping close around the wheels.

To further armor its sensitive components against heat, Boeing added fusible-mass tanks to the design: two small tubs of paraffin wax glued with a heat-conductive, silicone adhesive to the components governing the rover's drive controls and navigation, and another three tanks linked to the LCRU. Solid at the beginning of a drive, the wax would melt as it absorbed the excess heat it drew from the gear. At drive's end, it would cool and resolidify, then be ready to repeat its low-tech duty on the next outing.

The roll bar was ditched as an unnecessary and labor-intensive complication. Much about the crew station was nailed down, too. NASA had tested a mock-up of Boeing's seat design in a KC-135 Stratotanker, an air force variant of the Boeing 707 used for mid-air refueling. The big jet had carved a series of parabolic arcs in the sky, during which astronauts and NASA engineers experienced half-minute spells of simulated one-sixth gravity—and the crews hurried to get in and out of the seats, figure out their seat belts, and work out the positioning of the footrests, joystick, and other components. Such flights were the readiest way to simulate conditions in space; much of the equipment and procedures the astronauts used in the capsule and on spacewalks were likewise tested in a KC-135. They weren't comfortable, however. As described to me by those who participated in the ordeals, the first couple of parabolic arcs were usually pretty fun. Soon, though, the cumulative effects of shuddering climbs, near-weightless floats, steep dives, and crushing g-forces took their toll. "A lot of people didn't make it through the full series without calling on Rourke in New York," recalled astronaut Gerald "Jerry" Carr, who flew many times to check out rover hardware. Everyone called the plane the Vomit Comet.

Carr's flights had answered questions about how smoothly a crewman could get in and out of the seats in one-sixth gravity while wearing a space suit. Further flights prompted designers to slide the seats forward a few inches to better balance the rover, set the seat backs at a ten-degree lean, and fix the size and shape of handholds and toeholds. On one two-day series of flights, Carr completed fifty-nine parabolas, presumably achieving an unrivaled system cleanse. "I must have repressed that," he said when reminded of the feat a half century later. "The Vomit Comet was not a pleasant thing to do."

40

ALTHOUGH THE REVIEW WENT WELL, IN THE DAYS THAT FOLLOWED, Morea felt a growing unease. Three months into the project, Boeing was spending NASA's money on schedule, with its payroll actually below projections. But the schedule comprised a slew of small milestones, each rather insignificant on its own but part of a web of interdependent tasks on which meeting the final deadline relied. Morea's people, eagle eyed about those milestones, reported that Boeing and GM were missing some of them.

Work had fallen behind in areas that had nothing to do with the review's open action items or its few firm decisions. Subsystem design and testing should have been well under way by now, but wasn't. Morea went to see Henry Kudish at Boeing. "I said, 'Henry, you have to make these milestones,'" he recalled. "He said he understood." The company didn't seem worried about the slippage and ascribed most of it to NASA's ongoing tweaks to the rover's design. It would catch up.

Such talk brought Morea little comfort. It didn't much matter that Boeing's spending was on target if there was too little to show for it. Down the road, when the project got into the pricey fabrication stage, the contractors would need money they'd already spent. That meant cost overruns. Boeing assured him that wouldn't happen. Still, the disconnect between spending and productivity gnawed at him, as did Boeing's rather inscrutable project management, which made no sense to Morea. Granted, it was early, and Boeing and its subcontractors were still finding their feet, but he sensed that too many players were responsible for too many pieces of the project, and it wasn't clear they talked to one another.

Boeing's relationship with Sam Romano, Frank Pavlics, and the GM team in Santa Barbara was especially cloudy. In a perfect world, Morea wouldn't worry about it: NASA's contract was with Boeing, and it was up to Boeing to keep tabs on its subcontractors. But the rover's most important components were coming from GM, and Morea saw little to suggest that Boeing was staying on top of the work there—and he wasn't confident that the Santa Barbara shop knew how to manage this sort of NASA contract without the prime contractor's guidance. "The General Motors people, Pavlics and that whole group, these guys had always worked in a research and development mode," Morea told me. "They were good engineers, but they'd never done anything like this."

His worries were borne out when he met later with Romano and discovered that the Santa Barbara team was building a 1G trainer that lacked several changes that Boeing and the Marshall Center had signed off on the previous month—evidently, word of the changes had not passed from Boeing to GM. With the trainer scheduled for delivery in September 1970, there wasn't time to undo what was already done. Morea told Romano to proceed with the trainer as it was.

"My main message was we better establish a better communication system between the three affected parties, because it stinks," he wrote of the meeting in a memo to himself. With so little time, they'd blow it "in a system where GM makes assumptions, moves, and in the meantime we & Boeing are talking of a different system, etc." This communication gap, Morea believed, was exacerbated by the lack of a definitive contract between Boeing and GM. The companies were working off a letter contract that, in the manner of such documents, provided little detail about how they'd interact.

He soon had a lot more to worry about. By late February, when Morea met with Kudish to discuss Boeing's plan for catching up to the schedule, the company had missed so many milestones that additional slips were all but guaranteed. Every piece of the program, it seemed, was running behind. His conversation with the Boeing project chief unnerved him further. "We did not see even the slightest glimmer of a saleable recovery plan," his handwritten notes about the exchange read. "Major concern is the Boeing/GM relationship—it almost seems we have 2 separate contractors." He later described the meeting as a "fiasco."

In retrospect, Morea would come to feel great empathy for Kudish, whom he regarded as a good engineer, well-organized manager, and honorable man. "He understood there was a problem and he needed to do something about it," he told me. "But Kudish had practically no authority within the Boeing hierarchy. He had no muscle within the company. He was all by himself. 'Go make it happen,' the company tells him, and that was it. It was pretty clear, the more we worked with him, that he didn't have the resources within Boeing. If he needed help somewhere, it wasn't coming.

"You have to recognize, also, that the team GM had on this was a research group. They didn't have managerial experience with a program that was cost controlled and schedule controlled. They were concept people. So that was another problem that Kudish had: not only did he not have the support within his own management, but he had this group at GM that acted completely differently from what he was used to."

Regardless, the problems were Boeing's to fix. So a day after the "fiasco," Morea met again with Kudish and F. B. "Red" Williams, Boeing's Huntsville manager, this time laying down the law: in one week, Boeing had to give him both a clear program plan and a detailed timeline for catching up to the schedule, and he wanted evidence that the company was riding herd on all of its subcontractors. "It is my personal assessment that we have the workings of a 1–3 month slip in the delivery of the first flight article," he wrote in his weekly notes to Rees—who by now had assumed the Marshall Center's directorship— unless Boeing took "immediate management action . . . to pull up the schedule slippages indicated in numerous areas." To encourage that action, Morea organized a "tiger team" of Marshall program management and lab experts to work with the company and to "assess the total Boeing planning and capability to do the job."

41

A ROUTINE PART OF ANY MARSHALL PROJECT WAS "CONTRACTOR penetration." The center put one or more of its engineers inside a company's shop to monitor its progress, answer questions, and provide guidance, and to give NASA a realistic picture of where things stood. As Morea recalled, for instance, "Our guy would go down into the lab in Santa Barbara every day and see what was going on. If he sensed a problem, he'd go to Sam Romano." If a project appeared to be going south despite that on-site presence, the center's next step was a tiger team: as many as forty NASA engineers would flood the contractor's facilities to assess the state of affairs.

The team that Morea created in early March descended on Boeing Huntsville and the Santa Barbara lab. In two weeks, it found that management links between the companies were informal and weak. It found that Boeing's project manager was, in the words of one Marshall Center administrator, "running the project from his hip pocket." It found that the GM lab was unaccustomed to such a fast-moving program, had no scheduling controls in place, and that its quality control left much to be desired. As the weekly notes put it, "GM is the weak link in the LRV development and manufacturing cycle."

The tiger team's report prompted Rocco Petrone, NASA's Apollo program director, to telephone AC Electronics's headquarters in Milwaukee with word that its Santa Barbara operation was in trouble and to ask that the company overlay more experienced program managers on the staff there; in response, AC dispatched investigators to check out the situation. At the same time, Boeing's top management in Seattle started its own internal review, sending a team to Huntsville under the command of Oliver C. Boileau, the new vice president of its Aerospace Division.

That kind of attention was often enough to whip an ailing project into shape. As the Marshall Center waited and hoped for just that, contracts negotiator Mike Grant reported that Boeing's formal contract with GM was moving toward signatures. That would have been good news, except that the agreement appeared to have a built-in flaw: it promised a 4 percent minimum fee to the Santa Barbara lab, at the same time that Boeing's own fee from NASA was endangered.

By now, Morea could see that Boeing was moving too slowly on the project and too fast in its spending, which put it in danger of busting its deadline, its budget, or both. If the company didn't hurry to turn things around, it could well earn just the minimum fee from NASA and have to pay part or all of GM's fee from its own pocket. No government agency relishes the notion of partnering with a contractor that stands to lose money: it saps hustle, threatens quality, and never ends happily. Boeing's vulnerability, Grant warned, put NASA and the rover at risk.

42

BY EARLY APRIL, A LITTLE MORE THAN FIVE MONTHS IN, THERE WAS little joy in Huntsville. Boeing now calculated the rover's weight at 448.9 pounds. The vibration test unit, needed to verify the integrity of the final design, was running eighteen days late. And the company realized it was unprepared to conduct the critical design review planned for April 15—two months' worth of unprepared. The review was a key moment, a gathering at which the development stage of a project ends, and the design is frozen, so that blueprints can be drawn up and fabrication started. Boeing wanted it pushed back to June 16.

Morea's patience reached its limit. "I finally got frustrated enough to go over Kudish's head and go to Boeing and tell them, 'Look, this man needs some help,'" he said. But "instead of giving him help, they brought in a more senior manager who, they felt, could talk better for the company."

The hammer dropped in mid-April, when Boeing replaced its rover leadership team. Henry Kudish was out as project manager. In stepped forty-eight-year-old Earl Houtz, who'd come to work for Boeing in 1941, only to leave three years later to earn two Bronze Stars in the Philippines during World War II. Since his discharge, he had tackled a variety of NASA assignments for the company, including stints as a supervisor at Cape Kennedy and as a manager on the Saturn V.

"Earl was an old iron banger who had a lot of background in getting hardware out the door," Boeing's Al Haraway told me. "Both he and Henry were really good men. The LRV was not a big program, but everyone wanted it to succeed, and there are often casualties in those sorts of things." Kudish had marshaled the effort to land the contract for Boeing in the first place, Haraway noted. But "there's a big difference between bear catchers and bear skinners—they're just different people."

Herman Newman was gone, too, as the rover's chief engineer. In his place, Boeing appointed Eugene Cowart, who was already in Huntsville working on the company's missile programs. Tough and funny, the forty-six-year-old Louisiana native had flown forty-five missions over Europe as a B-26 bomber pilot. After the war, he earned a degree in mechanical engineering at Alabama Polytechnic Institute—now Auburn University—then worked in the oil industry and at a paper mill before he joined Boeing in 1956.

"I'd been sent to take over stuff before, and this time I had people say to me, 'Hell, I didn't know we were in so bad a shape that *you*'d show up,'" Cowart recalled when I visited him at his home in Huntsville in the summer of 2019. "But they *were* in trouble with the rover. They were late, and the money was just about gone.

"A lot of contracts are tough, and this one was especially tough," he said. "We were breaking ground in a lot of ways. We faced a lot of unknown unknowns—things we didn't know that we didn't know. But I'd flown forty-five missions, and I was a pretty hard cookie. I wanted to see if we could turn it around."

43

THE SHAKE-UP WENT UNNOTICED OUTSIDE HUNTSVILLE, OVER-shadowed by a far larger NASA drama. On April 13, 1970, Apollo 13 was on its way to the third planned lunar landing when an oxygen tank in the spacecraft exploded. Crippled, running out of air, and all but powerless, the ship seemed doomed, and its crew with it. In the face of one life-threatening setback after another, NASA got the men home. Its on-the-fly solutions testified to the agency's brains, poise, and creativity, but the emergency rekindled doubts in the public, press, and Congress about the wisdom of continuing moon flights. The Apollo program had already achieved its goals. Why was NASA putting its astronauts at needless risk, when it had reached the moon and beaten the Russians, who had all but scrapped any plans to land their own people there? America's luck was bound to run out.

At the Marshall Center, everyone tried to stay focused on the hardware that would truly revolutionize the moon visits to come, and to ignore the gnawing possibility that the program wouldn't last long enough to see it put to use. It didn't help that each week the LRV brought new exasperations. During a load test on the 1G trainer chassis in Santa Barbara, the frame bent and stayed that way. Rather than handle the situation in NASA style, with a thorough investigation to pin down the failure's exact cause, the GM team tried to straighten the chassis. "At the conclusion of this process, a weld failure was noted," Morea reported to Rees. "GM then apparently proceeded to try to repair the weld."

Nothing made the Marshall labs crazier than a breakdown in test discipline. The whole point of testing was to get answers, and now the trainer's chassis was ruined with no insight gleaned as to why. Morea jumped on Boeing, which in turn jumped on Santa Barbara, writing that it was "deeply concerned over the apparent lack of quality discipline within the AC/DRL facility," and that the "magnitude of the problem requires immediate top management attention."

As if that wasn't troubling enough, Boeing then wrote to the Marshall Center that it had several "areas of concern" about the contract's requirements, including the rover's weight and performance—as in, it wasn't sure it could meet the specs. Elbert "E. B." Craig, the project's contract officer at Marshall, fired off a letter criticizing the company's "patent reluctance" to speak forthrightly. He insisted that Boeing "reaffirm in writing your capability to comply with the requirements of the contract" or specify how it might fall short. Either way, Marshall needed a clear understanding of just what the rover would be able to do, and he needed it right away.

Craig was still waiting for an answer a month later, when Boeing's new Huntsville management admitted what Morea had figured out long before: the LRV project had torn through most of its budget. It would spend all of the contract's $18.6 million target cost in July, nine months before the first rover was due. Beyond that, the company was forecasting a "significant" overrun. Just how significant? Unclear. Boeing's projected totals ranged from $23 million to $29 million.

A hasty Marshall Center audit said otherwise. At a *minimum*, the cost at completion would be $29 million, and even that was a dream—it would require laying off 70 percent of the entire rover workforce over the next three months. Lee James, the center's head of program management, wrote to Rees that the dilemma was almost entirely Boeing's fault. But, he warned, as a percentage of the target cost, the overrun could be entering historic territory: it might draw comparisons with a recent air force contract for the C-5A Galaxy, a heavy-lift cargo jet on which Lockheed had overspent its budget by a billion dollars, sparking a congressional probe and a public scandal. In other words, the LRV project's financial woes "could quickly become the subject of outside interest." NASA, already under the budget gun, knew that if word of the overrun got out, the blame wouldn't be laid solely at Boeing's feet.

On Thursday, June 18, Morea convened a meeting in his office with the project's contracts team, a program control officer, and a NASA lawyer named M. X. Shanahan to review where they stood. It wasn't a good place. "Mr. Shanahan is particularly sensitive towards the possibility of Congressional activity initiated through the efforts of either Grumman or Bendix, both companies having been eliminated by selection of Boeing," Morea wrote in a memorandum for the record. And while that hadn't happened yet, "despite what appears to be an extraordinary degree of dissemination of that fact that Boeing is in trouble on the program," everyone at the meeting worried that it was only a matter of time.

A public scandal, bad as that would be, wasn't the only danger they faced. The overrun would erase Boeing's incentives, dooming the company to earn less than $200,000, no matter how well or badly it tackled the remaining work. "The contractor cannot be influenced at all . . . through the incentive formula," Morea observed. An arrangement intended to protect the government now threatened to work against it.

What followed was a do-or-die session in which Morea and his team weighed their options "at this very crucial point in the project," among which was scrapping the whole deal. They ticked through their choices, mulling the pros and cons of each. First, they could accuse Boeing of default and end the contract on those grounds. "Although a fairly good case could be made for such action," Morea summarized, it wasn't likely that Boeing would go down "without a vigorous fight"—claiming, for example, that the crazily tight schedule all but guaranteed failure—so a government win was no sure thing. More importantly, it would strike any chance of putting a rover on the moon before the Apollo program ended.

Option two: they could end the contract by claiming convenience, in essence saying the government had changed its mind about the project. "A legitimate alternative," Morea judged, but NASA would have to spend more money just to get out of the agreement. And once again, there'd be no rover.

They could invoke the contract's limitation-of-cost clause. This would fully fund the project right away, while cutting off any further cash. The move would force Boeing to either complete the job at its own expense (which Morea thought unlikely) or stop work—in which case, finishing the rover would fall on Marshall, which was already overworked and understaffed. Not a great choice.

They could also simply continue to dole out the public's money until the job was done. "Not an entirely unacceptable course of action," Morea concluded, "but recognized as an extremely difficult (and possibly dangerous) one. This action would offer the greatest invitation to outside criticism." It would also require that NASA really put the screws to Boeing and its subcontractors "to ensure that the projected

9–10 million dollar overrun does not grow to 12 or 15 million or more." One plus: it wouldn't disrupt the project, such as it was.

Finally, the group pondered converting the contract from cost-plus-incentive to some other form—firm fixed-price, perhaps, which would commit Boeing to a stated final price, beyond which the company would have to cover any overrun. Morea couldn't imagine that Boeing would agree to it. But, he wrote, there was "no better way for NASA to find out whether a 9–10 million dollar overrun estimate is realistic than to ask Boeing to accept full responsibility for any expenditure in excess of that amount." Furthermore, if it were facing the termination of the whole deal, Boeing might see a price ceiling as the lesser of two evils.

The meeting adjourned without a decision, though the arguments against most of the choices were so overwhelming that in retrospect, just one—staying the course—seemed viable. The fact that canceling the program was even on the table, however, reflects just how critical the situation had become.

It would get more so. Within days, Boeing upped its projected cost at completion to $29.7 million. With that, Eberhard Rees wrote to George H. Stoner, Boeing's executive vice president, to say that NASA was alarmed by the project's woes and mystified by the company's failure to get the job done at the agreed-upon price. "We would rather adhere to the concept of having the job done by Boeing without too much government interference," he wrote. But the rover had demonstrated that NASA had "to watch each and every step the contractor performs."

Stoner's reply was contrite. "I am particularly chagrined that you had to write me a letter in the nature of your June 23 communication," he wrote. "Your disappointment . . . cannot exceed mine." Oliver Boileau, the Aerospace Group's vice president, would henceforth personally oversee the project, Stoner assured him, "while applying every bit of ingenuity available to reducing the remaining cost and to assist you in every way we can think of to reduce the burden of [the Marshall

Center] in dealing with the total situation." Barring any unforeseen problems, "the current cost estimate of $29.7M to completion will be achieved."

Meanwhile, the critical design review took place, in two parts. They turned up just a few snags, but Boeing hinted again that it would not meet the rover's weight requirement. That prompted another letter from E. B. Craig, the contract officer, demanding that Boeing put it in writing. Again, he was kept waiting for a response.

44

IT WAS IN THESE EARLY, FITFUL DAYS OF SUMMER THAT MIECZYSLAW Gregory Bekker retired from General Motors at sixty-five. He did so with his legacy secure: just fourteen years had passed since his move to the United States, and in that time, he had founded an entirely new branch of engineering, influenced a generation of its practitioners, and laid much of the theoretical and practical foundation for lunar mobility.

Bekker's job title was senior scientist, and, as such, he had no underlings in any formal sense. But his imprint was on every vehicle the lab produced and informed the work of its competitors as well. Years ahead of his peers in his thinking, and more daring and imaginative in his designs, Greg Bekker might not have been the father of the rover—many shared its paternity—but it would not have happened without him.

"Perhaps I have been a precursor of some real progress in lunar surface locomotion," he understated in a letter years later. "In any case, I was closely associated with the work by NASA, in that respect."

He left behind a lab struggling for a handle on the rover job, especially a piece of gear that by all rights should have been the simplest part of the contract to fulfill. While the 1G trainer had to emulate all the behaviors the astronauts would experience on the actual rover, it wasn't bound for the moon, and could thus rely on down-to-earth parts and materials. In place of the harmonic drive, it used a traditional planetary-gear transmission. It didn't have to be folded, freeing GM to build a less complicated rigid frame. No seals against the hard vacuum of space were necessary. And it could weigh more than twice the rover's four hundred pounds and still acclimate the crews to driving the real thing.

But the trainer had sustained a chain of setbacks. The new chassis fell far behind schedule, then failed inspection and had to be reworked. Boeing Huntsville wrestled with myriad small but irksome complications on the trainer's crew station and instrument console. Soon the machine's delivery was pushed back two months, to November. That promised to cut into crew training time.

As the delays mounted, Don Beattie, who managed the Apollo Lunar Surface Experiments program, had put in a call to Rutledge "Putty" Mills, a mechanical engineer and automotive whiz who worked for the Geological Survey in Flagstaff. We're in a bind, Beattie told him. We need to get rover training started. Could you build something we could use?

Mills replied that he could. He met Beattie in Huntsville the following week and came home with Polaroids of the rovers under construction and a sheaf of Boeing's blueprints. "Of course, I had to make a simpler version," he recalled later. "There was no way I was going to make an identical vehicle in ninety days."

What he did build was the "Grover," for geologic rover. It matched the 1G trainer's dimensions and overall appearance but relied on creative fakery for the details. Mills couldn't find a drive system to mimic exactly the GM setup, but he did track down four landing gear motors from old B-26 bombers that came close enough. A surplus dealer in Los Angeles sold them to him for $12.50 apiece. Likewise, he didn't have time to custom build a torsion bar suspension, but he stumbled upon a reasonable facsimile in an unlikely place—inside Morris Minors, humble old British sedans heaped in Phoenix junkyards. "Here again," he said, "those things were really cheap."

The navigation system consisted of a $12 speedometer and an electric gyrocompass salvaged from a Frontier Airlines plane. That last part was likely the Grover's most expensive component: "It might have been that we had eight hundred bucks in it, or something like that," Mills said. "The rest of the Grover was built out of stock—you know, from our metal racks."

Assisted by two other USGS staffers, Mills built the mock trainer in a small shop on Flagstaff's east side, finishing the job by late July. Later, he added refinements to better mirror the actual rover—antennas, a rear-end tool rack, a fake lunar communications relay unit, and a real ground-control TV camera. The bill for the job was so low—$1,900—that Beattie, apparently aware that the real 1G trainer was on its way to costing hundreds of times as much, insisted on telling NASA that the Grover cost $20,000.

"Let me tell you, Putty Mills is a frickin' genius," Survey geologist Gerald G. Schaber told me. "The Grover looked a lot like the LRV, and it drove a lot like it. It did about seven miles per hour, which is what we needed it to do—that's about as fast as we expected the crews would be able to go on the moon."

The Grover would prove central to training exercises in the Southwest—in familiarizing the astronauts with the rover's controls, drilling them on spotting and avoiding craters, and using a vehicle on rock hunts. They drove it at the Cinder Lake crater fields, at the Merriam Crater outside of Flagstaff, and on the lands of the Navajo Nation. Mills hauled the Grover to the California desert, Nevada's Lunar Crater, and the Rio Grande Gorge near Taos, New Mexico.

Everywhere they trained with it sharpened not only their driving skills but their ability to pinpoint their location and follow a prescribed course. Practice boosted their speed and efficiency at pulling tools, gathering samples, and organizing what they'd found. They

Apollo 16 astronauts John Young, right, and Charlie Duke practice in the Grover near Taos, New Mexico. (USGS)

also worked out their own roles in the machine. "We didn't know how, at first, how we would portion out the duties of driving, riding, picture taking—all that stuff," recalled astronaut Charlie Duke, who with mission commander John Young was slated to put the real rover to its first use on Apollo 16. "What we decided is that on sixteen, John would be the driver, and I'd be the navigator. That was two distinct jobs. Since the handle was in between the seats, I could drive, and I practiced a little bit just so that if he got incapacitated, I could take over.

"Since you didn't have TV under way, and you could go a couple of kilometers between stops, I was sort of the travel guide as well as the navigator. As John drove, I was keeping him on course, but I was also taking pictures every fifty meters," using a custom-made, motorized Hasselblad attached to the chest of his space suit. "And I was describing the terrain: 'Now we're in a blocky area—blocks of such-and-such size,' and 'Now the rocks have disappeared, and we're going over another ridge, and it's smooth but undulating.' Those kinds of descriptions, so that they could match where we were to the pictures I took. It was kind of like being a travel guide on a tour bus."

The other crews decided on the same division of labor: the commander would drive the rover, and the lunar module pilot would ride. "Occasionally we turned on the movie camera," Duke added. "We had a sixteen-millimeter movie camera that was mounted to a post right next to the instrument panel." Its primary purpose was to capture moving pictures while the rover was itself moving. NASA called it a "data acquisition camera," or DAC. The label was so vague as to be virtually meaningless, but it gave the astronauts another acronym to use.

When the astronauts weren't on hand, Survey staffers used the Grover to time and streamline sample gathering for the upcoming Apollo missions. By monitoring its TV and radio feeds from their offices in Flagstaff, geologists practiced the role they'd play at Mission Control in real time. In terms of bang for the buck, the Grover might have been the Apollo program's best purchase.

Then again, if things had been running smoothly in Huntsville and Santa Barbara, it wouldn't have been necessary at all.

45

THE LRV PROJECT BEGAN AUGUST 1970 IN A PARTICULARLY PRE-carious state. The rover was now projected to weigh nearly five hundred pounds, and technical issues abounded. A malfunctioning shaker machine broke the vibration test unit. Hardware problems had pushed back assembly of the absolutely vital qualification test unit. This delay would force Boeing to undertake the risky business of running its final, all-systems tests of the rover even as it built the real deal.

The torsion bars in the rover's suspension were found to be made of a steel alloy susceptible to breakage and were rejected by Marshall's labs. Tests on the batteries had to be halted when flaws were discovered that required their redesign; the test schedule slipped by more than a month. One of the test units was running so far behind schedule that NASA decided it wouldn't be much good, and canceled it. The rover's electric steering motors were arriving at AC/DRL with design and quality flaws. Marshall pondered whether to pull the entire steering package in-house.

The harmonic drives were running late in delivery, too, and were plagued by quality and performance issues when they showed up. Even the one element of the rover that seemed completely figured out was running into trouble: the wheels. Doubts about GM's wire-mesh design had come up early in the year when astronauts who'd walked on the moon met again with NASA officials, along with Bekker and Pavlics. They compared lunar soil to a beach on Earth—loose on the surface, firmer and more cohesive with depth—and "felt pretty strongly that some type of lugs or cleats from one-half to one inch should be in the wheel design," Morea reported.

Bekker, one of the world's leading authorities on lugs and cleats, had listened quietly. He sensed that Neil Armstrong "was particularly skeptical in respect to the wire-mesh wheel developed at GM," he recalled later. When Bekker did speak up "to convince those present that it is all right," Buzz Aldrin sketched his own design for an alternative he reckoned might work. It featured seven hiking boots radiating from its hub. He suggested Bekker try something like that. "We will," Bekker deadpanned, "if you write an order." With that, Aldrin wrote one up and signed it, Bekker recalled, "amidst general laughter."

But it so happened that Armstrong's skepticism was shared by the Army Corps of Engineers, and that was no laughing matter. The Corps's Waterways Experiment Station near Vicksburg, Mississippi, was renowned mostly for its use of gigantic models to simulate the behavior of rivers and watersheds: the station had a working scale model of the entire Mississippi River basin that covered two hundred acres. Less celebrated was its study of soil mechanics, which years before had made it a not-always-friendly rival of Bekker's lab at the Detroit Arsenal. As part of its usual due diligence, the Marshall Center had turned to the Corps to test the wire-mesh design.

It found that in some soils the GM tire could not climb a twenty-five-degree slope, which NASA viewed as an absolute prerequisite. Covering the mesh in fabric improved its grip, but the Corps believed that both the open and closed versions fell short. For months since, GM and the army, assisted by engineers from the Marshall Center, had fiddled with variations in the tread pattern. The Corps remained unsatisfied.

Meanwhile, a three-day meeting in Santa Barbara brought to light another design issue, this one looming large in the minds of the astronauts. All through the summer, they and the Marshall Center's test subjects had donned space suits to check the ergonomics of the rover's crew station and had reached consensus that the joystick was almost impossible to use. The problem wasn't the stick itself but the astronauts' gloves: when pressurized, they were so stiff that closing them around the controller was exhausting. Within minutes, the act became downright painful.

There was no changing the gloves, so Houston mandated that Morea and company change the stick. Coming so late in the program, the switch had to be made fast. And it was—within days, Houston came up with a redesign, turning the vertical pistol grip into a short, T-shaped handle. Instead of closing his hand around it, the rover's driver need only rest his palm on the crossbar. A nudge would move it.

Astronauts Young and Duke went to Marshall to test a rough prototype. Each spent about two hours using it in a simulator before signing off on the change. The process of getting the new controller drawn up and fabricated passed to Boeing and GM.

46

AT NASA HEADQUARTERS, THESE TECHNICAL CHALLENGES TOOK A backseat to the project's deepening budget woes, which in the closing days of July prompted Dale Myers, the associate administrator, to suggest that the time had come to impose a cost ceiling on the contractors. Rees did his best to dissuade him. "I must admit to you there is some cost risk if no cost ceiling is imposed," he wrote to Myers. "However, if a cost ceiling is imposed, considerable schedule and quality risk would be inevitable." He concluded: "I believe we should complete our contract as planned."

He took this stand despite discouraging news from Morea. Boeing was still citing $29.7 million as its final cost, but that number did not include the design and hardware changes mandated by the almost inevitable test failures to come; the company would surely "break the $30 million mark soon." Headquarters kept the pressure high. On August 3 Myers asked Rees for a firm commitment on a final price.

"The problems of the LRV seem to be never ending," a frustrated Rees wrote Boeing's Oliver Boileau. In particular, he fumed, GM's performance was "most unsatisfactory," both in cost control and the quality of its hardware—and he had no reason to think Santa Barbara would "become responsive to our cost and schedule goals without affecting reliability."

Frank Pavlics was largely insulated from the bean counters at Boeing and NASA; his job required only that he build the best rover components he could in the time he had. His workdays were often brutally long, but they were taken up with problems he had always enjoyed solving. It was on the boss, Sam Romano, that the burdens of schedule and budget fell most heavily.

Romano had risen to many a past challenge. He was born to an Italian couple who'd joined the deeply conservative Plymouth

Brethren faith upon immigrating to the States; to describe his up-bringing as straitlaced, not to mention culturally confusing, is to understate matters dramatically. When World War II offered escape, he quit high school to join the navy. On his return, he earned his diploma before starting college.

In his midtwenties, Romano lost his first wife, Dolores Fitzsimons, to pneumonia, leaving him to raise their four-year-old son. He moved back in with his parents for a couple of years, during which he met a young secretary at the office, a fellow Garden Stater named Margaret "Marge" Jette. They married in 1957.

His early trials had toughened him. But no Santa Barbara job had brought the kind of professional pressure that Romano faced in the late summer of 1970. Until now, he had run a skunkworks, an idea factory. He and his engineers, all creative and talented, had always had the luxury of dreaming up solutions to difficult problems without having to put those ideas into production; that role belonged to others. Now they were obligated to actually build multiple copies of the most carefully engineered vehicle in history. The government's demand for perfection on each and every part was unyielding. NASA tested their work beyond common sense and rejected it until it passed.

In short, they were in over their heads. AC Electronics boss Paul Blasingame admitted as much in a letter to Rees: GM had failed "to properly define the system and the needs and desires of NASA," he wrote, especially when it came to the 1G trainer; the company had mistakenly believed that it could use its old six-wheeled designs as the basis for the machine, make a few quick changes, and deliver the thing on time. Obviously, it hadn't come close to meeting NASA's standards.

Adding to Romano's burden was that the space agency, like a middle school math teacher, expected its contractors to show their work. Its demand for documentation seemed to require keeping record of every sneeze in the GM lab. That had come as an unhappy surprise. "The requirements were just unreal," Frank Pavlics recalled.

"You had to write reports on just everything. If anything failed in testing, we had to write a detailed 'test failure' report. When we finished the job and delivered all the residual paperwork to NASA, all of the reports and all that, it took a truck to carry it all away."

Meeting the rover deadline meant ordering all hands to work on multiple shifts. This meant a lot of overtime, which boosted the shop's spending. There seemed no way to meet both timetable and budget. And to be fair to Romano and his team, there probably wasn't: whatever their failings, both GM and Boeing had undertaken a nearly impossible task. Quality lapses, fixed later, were a direct result of the project's frantic pace; time and careful attention to detail were inextricably linked. The ballooning cost had the same root. With the calmer schedule of a typical NASA undertaking, neither outfit would have needed to throw so many people at the job.

NASA came to this realization in mid-August, when the Saturn program's boss ordered an assessment of the LRV project management, under both Morea and the contractors. It found, as Morea wrote later, that "it really took the sort of effort Boeing and General Motors both had to put on this program to meet schedule, and that little or no padding or fat was evident." That insight did little to relieve the burden on everyone involved, however, especially Sam Romano. "They were just working around the clock," Marge Romano said. "He was meeting with all these different people and companies and dealing with them, and it was just conference calls all the time.

"He would get stressed. He would get very stressed," she said. "You could see how stressed out he was, and the worry about just trying to make it work. You could see it in his face."

Romano kept a diary of his days on the rover project, and while it doesn't provide much insight into his hopes and fears—he was an engineer, after all—it does chronicle the tasks and problems he juggled each day, rendered in a hurried scrawl that, as the summer progressed, grew to cover page after page. In late August came a development that surely added to the time he spent in meetings. AC Electronics, the Milwaukee-based GM division overseeing the lab, was combined with the carmaker's Delco Radio Division, based in Kokomo, Indiana, to form the Delco Electronics Division. In the musical chairs that accompanied any such consolidation, the Kokomo leadership came out on top. Santa Barbara now had a new layer of bosses.

At about the same time, Eberhard Rees finally responded to Dale Myers's August 3 request for a firm final price. He chose to do so by letter, rather than by teletypewriter, the usual mode of written communication between NASA's centers and headquarters in the days before email. "Because of the unusual sensitivity of cost numbers in the LRV program, I want to restrict exposure of such numbers to the absolute minimum," he explained. Then he justified his caution: Marshall had concluded that Boeing's latest cost estimate was "overly optimistic" and that a more realistic "outside top estimate" came to $34.7 million. "You have our commitment to this," he promised.

The Marshall director made this assurance while a major shift in the project was afoot. At an August 25 meeting in Rees's office, a Boeing delegation proposed transferring the rover's assembly and testing from Huntsville to the Boeing Space Center in Kent, Washington. It argued that the two-thousand-mile move might be mildly disruptive in the short term but would enable the company to finish the job with the help of 2,200 workers expert in manufacturing and quality control, versus just forty. They would also have access to more and better equipment, and would work just a shout from Boeing's test facilities. "We had schedule, performance, and weight issues," Boeing's Al Haraway told me. "So it was not inconceivable to get a little nervous as things progressed, and moving to Kent was a piece of the evidence.

"There was a desire to have additional management resources available as we went into the qualification test program. Boeing [was] more and more involved at a senior management level, and most of that senior management lived in Seattle."

Rees okayed the idea, pending a detailed plan from the company.

In the meantime, Sonny Morea wrote to Boeing's Earl Houtz with his "reservations and comments." His chief worries were loss or damage to small parts and tools during the move, as well as misplaced documentation and the errant delivery of gear already on order. Morea acknowledged the Kent facility would offer improvements over Huntsville, but he wanted to see more efficiency and "a commitment to reduce costs to the LRV Project and to the Saturn Program as a result of this move." Houtz replied that Boeing would save 7,700 man-hours of labor in Kent, which should keep costs from ballooning further—though it committed to nothing more. Morea was surprised by that, and not in a good way, but he approved the move anyhow.

In the meantime, NASA announced another momentous change. Two more Apollo missions were canceled.

47

BACK IN THE SPRING OF 1969, WHEN NASA HEADQUARTERS HAD given the go-ahead for the rover project, the agency contemplated as many as ten visits to the moon. The first, Apollo 11, would be a "G" mission—a simple landing, brief look around, and departure. Assuming it succeeded, the next four flights would be "H" missions, equipped to stay on the moon for up to two days. Their crewmen would make two extravehicular activities, or EVAs, on foot, each of several hours. While outside, they'd also set up an Apollo Lunar Surface Experiment Package, or ALSEP. An array of remotely monitored instruments powered by a small nuclear device, the package would transmit data about the moon's gravity, magnetism, seismic activity, and wispy atmosphere long after the astronauts left.

After the H missions, NASA had planned five longer "J" missions using the extended lunar module. Each would support a three-day visit and up to four extravehicular forays. Each would set up another experiment package and devote a great share of its time to scientific exploration. And the last four, Apollos 17 through 20, would carry a rover.

With Apollo 20's cancellation, the first rover deployment had moved up to Apollo 16, and its moonwalkers, John Young and Charlie Duke, had monitored the LRV's development with frequent visits to Huntsville and Santa Barbara. But on September 2, 1970, NASA's shrinking budget, coupled with Apollo 13's close call, spurred headquarters to further trim the lunar program. It canceled Apollos 15 and 19, leaving just four more moon flights.

Apollo 14 would now be the last H flight. The surviving missions would be renumbered: Apollo 15, originally planned as a walking H mission, would now be the first J and carry the first rover. Its landing crew, Dave Scott and Jim Irwin, would have to be trained in a hurry to drive on the moon. And only three LRVs, not four, would fly. "We'd now be the second users of the rover, and, of course, Apollo 17 was going to be the last," Duke said. "That was a disappointment, of course, but you have to expect those kinds of changes."

The Marshall Center had mulled canceling one of the flight rovers at least two months before, simply as a means to save money as the project blew through its budget. The inquiry had revealed that because the overrun had come in the development and testing of the rover's design, cutting one flight unit wouldn't save much: $700,000, or not enough to justify the trim. Now it had no choice in the matter. Rather than cancel the fourth rover outright, Marshall asked Boeing to build and test its components and subassemblies but leave them unbolted to the frame. LRV-4 would be a parts car for the other three.

In October, as truckloads of rover pieces made their way across the country to Kent, Marshall contract negotiator Mike Grant wrote a memo summarizing the project's chaotic first year. At its start, "Boeing and GM were motivated to make a profit—that's what business is all about," read the document, which Grant marked "Sensitive." "The government was motivated to spend the minimum amount of dollars to ensure an on-time delivery of a qualified LRV. The most superficial examination will reveal a similarity of the goals of the three parties. For in order for Boeing and GM to make a profit, they must align themselves with the goals of the government.

"Consider our current state of affairs, and I think we will find only one goal which is common to all three—delivery to the government of a qualified LRV in April," Grant wrote. "No longer is Boeing inherently interested in the control of cost for the purpose of profit. The contract incentive structure has effectively erased any motivation in this area." General Motors, which "did not fully comprehend, nor estimate for pricing purposes, the complexity and stringent requirements placed upon them by NASA," had little reason to be concerned about costs from the beginning, because its fee was guaranteed by Boeing. Besides, the company had "made it clear that they are not interested in future NASA business after this unhappy experience with LRV."

That left "only one principal party . . . vitally interested in cost control," Grant wrote, "and that is NASA." He concluded with a prediction: that the project would "continue to be a painfully trying task, for only the intangible motivating factor of 'corporate image' continues to operate in the government's favor."

As if on cue, Boeing submitted yet another cost revision. Citing a raft of technical issues, the company set the new price at $33.4 million, and the Marshall Center quickly established that design changes would add another half million or more to that. Oliver Boileau appealed to his Kent workforce to pull together. "Lunar Rover is a vital part of the national Apollo program," he wrote in a company bulletin. "Our ability to successfully perform can have a significant impact upon future space-oriented business for The Boeing Company. All Aerospace Group organizations must support this effort with the resources and cost control necessary to ensure successful performance."

48

THE PROJECT WAS COMING BACK TOGETHER IN KENT WHEN THE news arrived that NASA's rover would not be the first on the moon. On November 17, 1970, an unmanned Soviet lander, Luna 17, set down on the Sea of Rains. Ramps deployed from the craft, and down rolled an eight-wheeled, robotic machine called Lunokhod-1— Russian for "moonwalker."

It had been in the works since 1963. Like their American counterparts, Soviet engineers and scientists studying lunar mobility had been stymied by their ignorance of the lunar surface. Like U.S. contractors, they briefly considered walking, jumping, and creeping designs before they settled on wheels and tracks. Likewise, their Lunokhod underwent several evolutions before its final design emerged. The Soviets tested their prototypes on a mock moonscape on the Crimean Peninsula. And the Lunokhod's assembly had involved an effort just as frenzied as that at Boeing and GM: to deliver the first flight vehicle, Russian engineers often slept in their labs.

The robot's running gear had something in common with the rover's, too. Its chassis was built of aluminum and titanium. Its suspension relied on torsion bars. Its wheels, twenty inches in diameter and just shy of eight inches wide, were independently powered by small electric motors built into their hubs and rode on rims covered with steel wire mesh.

Resemblances ended there. Lunokhod's superstructure was a round magnesium tub topped with a hinged convex lid that opened to expose solar panels and cooling radiators. Stuffed into the tub were the electronics for two TV cameras, a high-resolution camera, radiation detectors, an X-ray spectrometer to chemically analyze the lunar regolith, and an extendable probe that measured its density. Ungainly but sturdy, the craft was just over seven feet long, a little more than four feet tall, and weighed 1,667 pounds. It moved at a crawl.

And it worked. After rolling off its lander, Lunokhod-1 set off across the moon's broken surface, operating during the two-week lunar days and hibernating through the equally long nights. It tested twenty-five soil samples and deployed its probes five hundred times. It snapped more than twenty thousand low-resolution photos and two hundred high-resolution panoramas. It was expected to survive for three months. It lasted more than nine. Before it finally died, the Soviet robot covered more than six and a half miles.

Early in the mission, and long before anyone understood how successful Lunokhod-1 would be, the Marshall Center went fishing for whatever lessons the Soviets had to share. George M. Low, NASA's acting administrator, was due to attend a conference with Soviet space officials and scientists in January 1971. Huntsville offered a list of questions Low might ask. "We probably would not change LRV hardware, even if we were to receive data points from Russia different from those in our design," the teletypewritten memo read, stressing that a response was "by no means considered essential to the currently planned LRV missions." With more accurate data, however, rover engineers "could optimize our mission energy consumption plan and thus derive more scientific benefits from Apollo 15." In exchange, the Marshall Center offered to share its own research.

Low's visit yielded a joint communique from the two governments, in which they pledged to exchange lunar soil samples and cooperate in space, weather, and environmental research, among other things. But, really, any findings the Soviets might have offered on the rover would have come too late to be of much help.

49

AS EXPECTED, THE ROVER PROJECT STUMBLED AFTER THE TRANSfer to Kent. All through November, it was plagued by continuing flaws in workmanship and late component deliveries. Problems persisted on

Frank Pavlics, left, and Sam Romano pose with their seemingly endless headache, the 1G trainer, on the day it finally rolled out of the Santa Barbara lab. (FACULTY OF AUTOMOTIVE AND CONSTRUCTION MACHINERY ENGINEERING, WARSAW UNIVERSITY OF TECHNOLOGY)

the 1G trainer. Testing was delayed on the first flight chassis because of missing paperwork. The qualification test unit, already running late, continued to slip. Barring a miracle, testing wouldn't be finished until mid-February 1971—more than a month after assembly needed to start on the first flight-ready rover to make the deadline for Apollo 15.

Rees decided it was time for another talk with Boeing's Stoner. In a December 3 telephone call, followed up with a letter the next day, the Marshall chief gave the Boeing boss an earful. The delays and failures irritated him for their own sake, but also because they threatened to push costs beyond $34 million. Lines of communication within Boeing were confusing—too many cooks provided too many points of entry for decisions that could push costs even higher. And the 1G trainer: Would it ever be finished?

Stoner huddled with his top rover people. They assured him that the move to Kent had boosted their confidence. They had a what-if plan in place to cope with unexpected trouble, and an additional shift of workers to get ahead of the test schedule. They also felt good about the hardware now coming from GM. The harmonic drives, in particular, seemed to be working as advertised. Stoner channeled their optimism in his reply to Rees. "The $34 million cost at completion provided to NASA in October can still be met unless we have major surprises out of the qual test program," he wrote. "We recommend that you remain committed to Rover for Apollo 15 based on our confidence that the Flight Unit delivery date of April 1 will be met."

And look: Within the week, the snags did, in fact, appear to be smoothing out. Chassis repairs were completed on the 1G trainer, and its new hand controller breezed through testing. On a day of celebration in Santa Barbara, Greg Bekker, Sonny Morea, astronaut Charlie Duke, and Boeing officials were on hand to watch the finished trainer roll into the parking lot. "I was the first one who drove the training vehicle," Pavlics told me. "It was a very good ride. The soft-spring wheels and the suspension gave it a very comfortable, good ride." They all posed for pictures, and, in short order, the trainer was delivered to Houston, where the astronauts drove it on a small proving ground nicknamed the Rock Pile. "The course requires considerable steering activity," Morea reported to Rees. "The astronauts have reacted favorably to the vehicle and to its simplicity of operation."

In the wake of myriad delays, the qualification test unit was finished in mid-December, and testing began at Kent the next day under the watchful eye of Marshall engineers. Now everything suddenly started going right. Over the first two weeks of January 1971, Boeing began assembling the first flight unit and threw more than two hundred additional workers on the job to speed it along. The company completed the chassis and was busy with the subassemblies. GM's hardware was on its way.

The extra manpower came at a price. When Boeing updated its runout cost for the program, it stood at $36.5 million. Morea asked for data to support the figure. In the time it took Boeing to gather it, the estimate jumped to $37.8 million—more than twice the contract's original target. "The cost forecast has spiraled upwards faster than we can react," the Marshall Center's contracts office complained. Morea's people hurried through a new round of negotiations and cemented the final price at $38.1 million.

The subject was not closed, however. In January 1971 the *Washington Post* broke the news that the rover project had far exceeded its contract price. Its story, headlined "Moon Runabout's Costs Run Away," was based on months-old estimates, putting the cost at $31 million. NASA had paid it "without fuss," the story said, "because it would rather have an expensive lunar rover on time than one that arrived late." The story left no doubt about the project's centrality to the last three lunar missions: Rocco Petrone, the head of the Apollo program, told the *Post* that while he was confident the rover would be ready for Apollo 15's July launch, the agency would delay the flight by a month, if necessary, to get the machine aboard.

The overrun caused a stir nonetheless. Staffers for the Senate subcommittee overseeing the space program wanted an explanation. What's more, there was talk that Congress might call in the U.S. General Accounting Office, the federal government's independent, nonpartisan auditing agency, to look at the books. Ben Milwitzky, who ran the Apollo Lunar Exploration Office at headquarters, told Morea that Petrone wanted a full rundown on the situation, so that he could brief Capitol Hill. "I gave Rocco Petrone that presentation," Morea said. "I tried to show that we had tried to protect the government by structuring the contract so that if there was a substantial cost overrun, the contractor got no fee. If he didn't deliver a car, he got no fee. If the car didn't work, he got no fee."

Morea did not care for Petrone, who would later become the Marshall Center's director and, some old-timers say, make it his mission to dismantle the organization and culture von Braun had built there. "When I presented this to Rocco Petrone, he was busy squealing," Morea told me. "He kept saying, 'Why would we penalize the contractor that way?' I tried to explain that the incentives were there to protect the government." The presentation also emphasized that while the cost had exceeded the contracted price, it promised to come very close to NASA's original ballpark figure for the project. Von Braun's initial authorization from headquarters had said the project couldn't top $40 million. And short of disaster, it wouldn't.

Petrone made his presentation. Morea waited to hear when to expect the GAO's arrival. "But I never heard anything after that," he said. "I was just waiting for these guys to come and terminate the program. So I called Ben Milwitzky and said, 'Ben, when is the GAO coming down here? We know they're coming, but it would be nice to know when.' He said, 'No one called you? They're not coming down.' It turned out that when Rocco Petrone gave his presentation, the congressional people were so impressed that there was a contract office down here concerned about the taxpayer that they called the GAO off of it."

Morea scowled. "And I never heard from Rocco Petrone about that."

50

COST ASIDE, MOREA'S PROBLEMS GREW MORE MINOR AND MANageable by the day. Most involved simple, quick fixes. Workers accidentally snapped a couple of the torsion bars in LRV-1's suspension. A wax heat sink on the rover's front end failed a pressure test and had to be redesigned. The steering motors failed late in their qualification testing, but it turned out they'd been jostled far more than they would be on a mission. Huntsville and Kent ran experiments to ensure the rover's electronics didn't interfere with those of the lunar module or with the wiring in the astronauts' backpacks.

With the deadline looming, NASA found itself adopting an uncharacteristic pragmatism about a few remaining snags. The Army Corps of Engineers continued to raise questions about the rover's wire-mesh wheels. GM had subjected them to its own testing with a mule, a one-seat dune buggy fitted with the rover's drivetrain and wheels that astronauts Jerry Carr and Jack Lousma drove up and down the dunes at Pismo Beach. Those runs, and further tests at the lab, revealed that the wheels' aluminum treads disintegrated after traveling the equivalent of roughly 112 miles—but that treads made of titanium did not. GM promptly made the switch.

Many of the Corps's remaining complaints had been addressed with a simple fix: using the tread strips to cover 50 percent of the wheel's contact area—down from as much as 80 percent—offered better traction. Even so, as late as March 1971, the Corps warned that the wheels' rolling resistance might chew into the rover's power supply and limit its speed. Morea was willing to live with that. In his view, the wheels were ready for Apollo 15 as they were.

Likewise, vibration tests had revealed that the folded rover's wheels chattered against the inside of quadrant 1 in the first minutes after launch. The lander's skin was lightweight and fragile; the metal-on-metal rubbing could be dangerous. The surest solution was to add a bumper to the lunar module to protect it. But Houston was so conscious of every ounce of extra weight, it decided against a fix. And as it happened, when Grumman later tested the lander and rover on Long Island, the problem disappeared.

A final worry occupied the Marshall Center now: the gear that released the LRV from the lunar module and lowered it to the ground. Even if the rover itself did come together in time for its July launch, the collective effort of hundreds of people—not to mention many millions of dollars—would go for naught if the astronauts couldn't actually disengage it from the lander and drive it away.

The contraption had to work with the reliability of an expensive watch every time. It had to happen quickly and require little muscle from the astronauts, for time and effort were precious. And it had to hold the rover tight until it was deployed. If the rover came loose during the violence of launch, it might entangle the lunar module in its shroud at the top of the Saturn's third stage, effectively killing the mission. If it sprang free while the module was descending to the lunar surface, it could kill the two men inside.

So the deployment gear—which went by the imprecise name space support equipment, or SSE—had to be as well engineered as the rover itself. Boeing's original concept was largely automatic: a single pull on a handle would set off springs, pulleys, and levers to snap open the tightly folded rover and deposit it on the regolith. From the start, this proved a headache. All through the spring of 1970, Boeing had struggled with the design, at one point growing so concerned with its reliability that it junked what it had and started over. The new approach also suffered "design deficiencies," as the company noted in a report that summer.

Boeing's Gene Cowart was present when those "deficiencies" became apparent. "We put on a big demonstration one day, over here in Huntsville," he told me. "Beforehand we said, 'You know, this is on the moon. We'd best simulate it. There'll be moondust everywhere.' So we went into town and got some very inexpensive talcum power. We had a lot of it. We sprinkled it everywhere." The demonstration included a lunar module mock-up and a suited astronaut deploying the rover from its ladder.

"The big day arrives, and Dr. Rees came," Cowart said. "The astronaut comes out and pulls the lever, and the rover went about halfway out and *clang*! Stops dead. Dust everywhere. And old man Rees said, 'I didn't think it was going to work like *that*!'"

Boeing again redesigned the system, which took most of July and August and added an unwelcome nineteen pounds to the project's weight. The new semiautomatic version required the astronauts to perform three separate but simple operations. After eight successful days of testing, the company professed its satisfaction with the machinery. Morea was unconvinced, however, reporting to Rees that the tests "do not give us the confidence we need to proceed with this design." Yet another iteration was shipped to Grumman for testing. After several days of dry runs, Morea could report that at last the deployment gear seemed to operate consistently. When the Marshall labs stepped in to evaluate it, they decided otherwise. Boeing went back to work.

Amid all this testing and redesign, Rees revealed that he was suffering "sleepless nights" over the rover's deployment. In an October 26, 1970, memo, he wondered "whether anybody has gone systematically through the whole timeline of environments the lunar rover has to go through," especially the landing. "Could a harder landing on the moon impair the LRV deployment?" he asked. "As I understand, Apollos 11 and 12's landings were really soft, but a harder landing like this could be envisioned." He also wanted to know "under what surface roughness condition (rocks, etc.) would a deployment be impossible," and asked that the labs "look into all these points if it has not been done already."

They were already on it. Anticipating that the lunar module might come to rest at a tilt, the next round of tests, in December 1970, called for two complete rehearsals of the system with a mock-up lander pitched at five different angles—perfectly level, and leaning 14.5 degrees forward, backward, and to the right and left. Three days into this testing, the saddle linking the rover's front end to the bottom of the deployment gear—a hinge that swung the machine out and away from the module—developed a crack. Boeing rushed to design a beefed-up version, but it broke, too. In mid-January 1971, with the project's deadline less than three months away, Boeing had a redesigned saddle at the ready and raced it into testing with the semi-automatic deployment system.

Hedging its bets, NASA ordered a fully manual backup system. The Marshall labs pitched in on the project. By late in the month, Morea could report that the resulting manual version had "shown highly promising results in prototype development tests.

"At this point," he wrote, "it appears that the manual design can be developed and qualified quicker than the automatic. Therefore, [Boeing] has been given technical direction to place their emphasis on the manual configuration." The Marshall Center did its own testing of the simplified gear and succeeded with sixty straight practice deployments, some of them with the lunar module mock-up heeled over hard. Astronauts who checked it out liked it. Morea instructed Boeing to halt its fabrication of the semiautomatic units.

But simplified did not mean simple, exactly. The space support equipment was a web of braces, support arms, ratcheting spools, pins, cables, and "deployment tapes" wound around braked reels, all fitted to attachment points within the lander's pie-shaped quadrant 1. The folded rover was mounted wheels in and belly out, so that the underside of its floorboard formed a hard shield for the rest of it. An insulating blanket was draped over the floorboard and held fast with Velcro straps.

Once out of the lander, the astronauts would first peel away this blanket. Then, after giving the deployment gear's latches and hinges a once-over, one of them would climb back up the ladder to pull a D-ring high on the lunar module's descent stage, releasing the mechanisms locking the rover in place. Its upper edge swung four degrees outward.

The astronaut on the ground held taut a deployment cable—basically a safety brake on the rover's movements—while his partner, stepping back off the ladder, pulled on a deployment tape on the quadrant's right side. As he pulled, the rover would slowly rotate out and down, like a drawbridge. At about forty-five degrees, its aft section flipped into position, its rear wheels sprang loose, and all locked into place. With further pulling, the front section and wheels unfolded. Pulling a second deployment tape on the left side of the quadrant lowered the machine to the lunar surface.

At that, the astronauts had only to disconnect the saddle from the rover's front end, unfold the seats, extend the fenders, lift the instrument console into position, and remove a pair of triangular braces from the sides of the frame. Pieces of those braces, fitted back into the chassis, became toeholds. Because the unfolded rover faced the lander, the astronauts had to push it a few yards backward before firing it up; Houston wanted them to start out driving forward rather than in reverse, and they needed room to turn around. But then, the rover was so light that they could pick it up and turn it around by hand.

The deployment instruction manual was a study in exactitude, with every step of the process described in painstaking detail. How to stand on the ladder. Which hand to use when pulling pins. Where to stand while the rover was unfolding—and where not to, in case the machine fell. And NASA did not rely on written instructions and diagrams alone. The astronauts rehearsed the deployment time and again. In shirtsleeves. In pressurized space suits. Over and over, until it required little thought at all.

51

ALL THE WHILE, THE APOLLO CREWS CONTINUED TO PRACTICE driving the rover. Out in the Desert Southwest, they trained in the Grover. At Houston, and later at Cape Kennedy, they drove the 1G trainer. "Occasionally we'd use it at the Cape, in our space suits," Charlie Duke recalled. "We had a special rover track bulldozed out in the wild area of the Cape.

"Wasn't really the same as the actual rover, though, because you're a lot heavier," he told me. "The steering wasn't nearly as sensitive, and it never bounced like the lunar version." To better simulate what they should expect from the genuine article, the Manned Spacecraft Center put the trainer in a harness that supported five-sixths of its weight. This partial gravity simulator, nicknamed Pogo, caused the machine to bounce convincingly over a mat of simulated knolls and craters, though the real lunar surface turned out to be quite a bit rougher.

When in Huntsville, the astronauts also used the Marshall Center's simulator. Part of the Computation Lab, the device had two main parts. The first, a dimly lit cubicle on hydraulic legs, contained a mock-up of the rover's crew station and featured a large, round video monitor displaying a driver's-eye view of the moon. Or so it appeared:

in reality, it was a foam rubber and latex model of the lunar surface, twenty-seven feet long, twelve feet wide, and mounted on rollers to simulate movement across its surface. Resting lightly on this rubber moonscape, a sensor with a TV camera rode the model's craters and ridges on four little balls; the sensor was linked to the hand controller and video screen in the crew station. The hydraulic legs gave the crew station six degrees of freedom—it could roll, pitch, or yaw beneath the driver in any combination—so that any bumps the sensor encountered, any hills it climbed, any tilts it made were replicated in real time in the mock rover.

Craig Sumner, then a twenty-year-old college intern, worked as a simulator test subject. Outfitted in a pressurized suit, he broke in the equipment and helped NASA refine its procedures for driving on another celestial body. "If you sat there in the simulator, looking straight ahead, and you compared what you saw with the film we later got from the LRV, you'd be amazed at how similar they were," Sumner told me. "The hand controller behaved in such a way that it replicated what it was truly like to drive it on the moon."

Finally, the astronauts trained aboard the Vomit Comet. During the plane's brief swings through weightlessness, they practiced getting in and out of the rover's seats in their pressurized suits. "That was good training," Sumner told me. "These guys would just throw themselves into the seats because the suits didn't bend easily. They'd grab a strap to force themselves into a seated position, then strap the seat belts to stay in."

They also practiced mounting the lunar communications relay unit and TV camera on the rover's front end, along with the antennas that would keep them in contact with the ground. When buzzers sounded a warning that the plane was starting its dive at the end of one-sixth gravity, Duke recalled, technicians helped him stretch out on the cabin floor, then hoisted him back to his feet as it neared the peak of its next climb. Encased in two hundred pounds of gear, he would have been knocked flat by the gravitational forces at the bottom of the parabola.

Technicians at the Kennedy Space Center inspect the newly arrived LRV-1, March 1971. (NASA)

52

BOEING AND MARSHALL CENTER ENGINEERS FINISHED TESTING the qualification unit in late February. By that time, LRV-1 was assembled and ready. The tests prompted tweaks to the flight-ready rover, but they were few. A last check of the crew station, with a suited astronaut at the rover's controls, yielded only a slight change to the seat belts. The government took formal possession of the finished LRV-1 on March 10, 1971. It flew the rover to the Kennedy Space Center five days later. Despite the blown deadlines, cost overruns, and anguish of the preceding seventeen months, the first rover was delivered two weeks early.

It was nowhere near NASA's 400-pound target for both rover and SSE, however. In fact, it missed it by nearly 25 percent—LRV-1 alone weighed just over 464 pounds, and its deployment gear another thirty. The excess would, by Houston's reckoning, cost the lunar module nearly ten seconds of hover, raising the stakes for the rover's debut. As if they weren't high enough.

Morea hurried to wrap up the few remaining items on his preflight agenda, among them the fenders. Built of a polyester-fiberglass blend, each consisted of a large section that bolted to the hub and curved close around the wheel, and a piece that slid back and forth to shorten or extend its coverage. The sliding extension was pulled back to its shortest setting for the flight, so that the fenders didn't get in the way when the rover was folded.

The design seemed ready for duty until, during a thermal-vacuum test in February, a set of the fenders *melted*. The Marshall labs rushed to ready a new set of slightly different construction. Those survived baking, but when Marshall engineers tried out their sliding extensions, they encountered a snag: "The cold temperature test at -105 degrees F resulted in an erratic sliding operation," the testers reported. "The hot temperature test at +185 degrees F resulted in a very smooth fender deployment action."

By now, however, with the rover at Cape Kennedy and no time left on the clock, the center's normally fastidious labs were forced to work around the problem. The fenders were fine as they were, they decided. The cold-weather balkiness could be fixed "by parking in the sun for a few minutes." The new fenders would be the rover's one splash of color, save for the foil shrouding its TV camera—they were orange, and all four bore U.S. flag decals, thereby satisfying a request from Dave Scott that the machine somewhere display the Stars and Stripes. The fenders were bolted onto LRV-1, and once a few additional replacement parts were added, and a handful of last-chance tests performed at the Cape, the rover was folded up and inserted into *Falcon*, Apollo 15's lunar module, on April 25, 1971.

Just one more task was left to perform. The rover was installed in the lunar module without its batteries, lest they lose their charge during the months before launch. A few days before that liftoff, a

Boeing crew took the batteries to the pad, carried them up the launch tower, and slipped through an access door into the shroud at the top of the Saturn's third stage. Inside was *Falcon*. The contractors removed the top half of the rover's thermal blanket and disconnected its floor panel, which gave them access to the folded-over front chassis and the battery mounts. Using a special sling and tethers to keep anything from dropping down into the rocket, they installed the batteries and hooked them via cable to monitoring equipment outside. Eighteen hours before launch, with the batteries showing a full charge, the crew returned to the lander, unhooked the cable, and reconnected the rover's floor panel and thermal shield.

The job was finished. The rover was ready.

PART SIX

ACROSS THE AIRLESS WILDS

53

WHEN APOLLO 14 CARRIED ALAN SHEPARD AND EDGAR MITCHELL to the moon for the last H mission in late January 1971, it was met with a weeklong yawn by the press and the public. The TV networks carried live coverage of the launch and paid some mind to the long translunar coast, where misfortune had called on Apollo 13, but offered only occasional snippets of the crew's time on the lunar surface. News from Fra Mauro was dutifully passed along by the country's newspapers, but with a palpable lack of enthusiasm. America had grown jaded about moon missions, it seemed, and on just the third to land.

But fact was, the H missions made for lousy TV. To viewers, they looked pretty much the same as Apollo 11: the astronauts were on foot. The TV camera was stationary and positioned near the lunar module. Its low-resolution footage captured the crew bouncing around base camp, doing nothing especially stirring. Apollo 14's stretches of real drama—when Shepard and Mitchell wandered the hills, desperately seeking Cone Crater—unfolded a half mile away, and unwitnessed. In fact, the mission's single made-for-TV moment is the only thing many older Americans remember about it: when Shepard attached the head of a 6-iron he'd smuggled aboard *Antares* to a lunar excavation tool and used it to whack a golf ball, crowing that it went "miles and miles and miles."

This time was different.

Weeks ahead of time, even months, it was plain that Apollo 15 was another kind of mission. Newspapers reported about it with renewed excitement: it would be longer, three days, with three ventures outside the lander. These astronauts would be real explorers, engaged in real science. A much-improved TV camera would follow them wherever they went. And, most important of all, they would have a car.

Journalists couldn't stop talking about the rover. They were fond of calling it a "moon buggy," to the chagrin of the Marshall Space Flight Center, but they seemed to have a genuine hunger for details about its design, performance, and potential contributions to lunar science. They wrote about its sky-high cost: by launch day, the press seemed to reach consensus that each flight-ready example represented an $8 million investment—a figure that involved some magical accounting, as it fell short of reality by nearly $5 million. Even at the lowball price, some stories called it the most expensive car ever made, the more so because it would be abandoned after just three days of driving.

The upshot is that Apollo 15 got a lot of attention, and much of what set it apart was the rover. Papers across the country illustrated their coverage of Saturday, July 31, 1971, the morning after the lunar module *Falcon* descended to the surface, with a detailed diagram of the rover's deployment from the lander. That same morning, scores more published an artist's rendition of the rover conquering the lurrain. The *New York Times* devoted nearly a full page to the machine. The networks began their live coverage shortly before the crew deployed the rover, followed its travels for hours, and did the same for its second venture from base.

Even advertisers piggybacked on rover fever. Consider the ad placed in the *Daily Leader* of Pontiac, Illinois, six days before the landing. "Apollo 15 will carry a Moon Buggy, and this Buggy has a color TV camera," it read. "Our TV screen will seem like a window on the Moon Buggy. To get a good view of the moon, you should have a nice clear window. A new RCA color TV from Schlosser's may be what you need. Why not be prepared for a thrilling ride on the moon?"

At 9:34 A.M. E.D.T. on July 26, 1971, Apollo 15 rose on a pillar of flame from its pad at the Cape and climbed with gathering speed past the crown of its launch tower. The sound of it took seconds to reach the nearest spectators, three miles away—a deep, almost seismic, roar, punctuated by a dry crackle of booms, each delivering a punch to the chest. Looking on were Wernher von Braun, Greg Bekker, Frank Pavlics, and Sam Romano.

The setting for the rover's maiden voyage was well chosen: the Hadley-Apennine region, an undulating plain rimmed on two sides by mountains the size of Everest and on a third by a canyon a mile wide and nearly a thousand feet deep. In place of the featureless gray flatlands that TV viewers had seen in past Apollo coverage was a moonscape of extremes, a backdrop without earthly equivalent.

Astronauts Dave Scott and Jim Irwin landed a mile from the canyon and three from the nearest mountain. A short time later, during a "stand-up EVA" from the lunar module's top hatch, Scott took a look around. One aspect of the place leapt out: it was relatively free of boulders and other obstructions. "Trafficability looks pretty good," he told Houston. "It's hummocky—I think we'll have to keep track of our position. But I think we can manipulate the rover fairly well in a straight line."

And so, the next morning, America watched over breakfast as the astronauts stepped onto the plain—more formally, and ominously, known as the Marsh of Decay—and released the mission's star attraction from quadrant 1. Frank Pavlics and Sam Romano, at Mission Control as technical advisors, saw the rover unfold and drop to the surface. "We were all cheering," Pavlics recalled. "I was sitting on needles. It was satisfying, and very exciting, to see the thing working like that."

But then the moon buggy hit a few snags. It hung up briefly on the saddle linking it to the lander. The astronauts twisted and tugged until it came loose. A few minutes later, as Scott unfolded his seat, the Velcro holding it in place proved stronger than lunar gravity, and he almost yanked the entire vehicle off the ground. Then, as Irwin filmed him, Scott climbed aboard, eased the machine forward, and discovered that he couldn't turn his front wheels. Irwin confirmed the bad news. "Got just rear steering, Dave."

Their capsule communicator, or capcom—a fellow astronaut named Joe Allen, through whom Mission Control routed all messages to and from the crew—suggested a number of switch combinations on the control panel. "Still no forward steering," Scott told him.

In Houston, Romano sat horror-struck. "I was sitting in the Mission Control Center, in the third row," he told a documentary filmmaker many years later. "Dr. von Braun was in the fifth row. So when they said, 'The front wheels are not steering,' my God, I was very, very nervous. The back of my neck began to swell, get red. My ears were red."

Pavlics, watching nearby, thought he knew what was up. A tiny component in the steering electronics, called a potentiometer, was failing to carry an electrical signal. "Sometimes the metal-to-metal connection in the potentiometer didn't make contact, and that was happening there," he told me. "So I suggested that they should exercise the hand controller, to move the two pieces together."

On the Hadley Plain, Dave Scott waited for instructions. Would Houston scrub using the rover? If so, Apollo 15 was going to be far less ambitious than everyone had hoped. Steering with just the rear wheels was certainly doable, and Mission Control's own rules for Apollo 15 allowed it, but NASA typically insisted on redundancy. Seconds passed before Allen told him, "Press on." Pavlics's fix would wait; the clock was running. Scott and Irwin hurried to attach the lunar communications relay unit, TV camera, and antennas and to load tools into the storage pallet behind their seats.

Irwin climbed on. "You really sit high," he said. "It's almost like standing up." He reached for his maps and found that his pressurized suit bent so little that he couldn't stretch that far. Minus Earth's gravity and atmosphere, their suits were fatter and stiffer, and the astronauts' lighter weight gave them less leverage. He fumbled with his seat belt. It wouldn't close. "I think it's too short, Dave."

Scott came around the rover to help. "Yeah, sure is."

"Don't waste time on it," Irwin told him. "I'll just hang on."

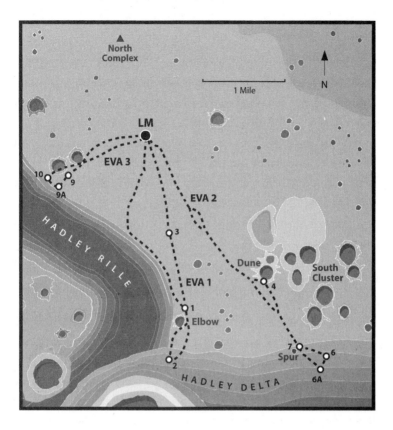

Apollo 15's travels in the wild, mountainous Hadley-Apennine region, a fitting stage for the rover's debut. (KEVIN SWIFT)

"No," Scott said, making perhaps his wisest decision of the day. "We can't lose you now. Got too far to go." He got the belt latched, climbed back in, and buckled his own belt. "Okay, Jim, here we go."

"Okay," Irwin said, "we're moving forward."

"Whew!" Scott said. "Hang on!"

54

SO BEGAN THE ADVENTURES OF LRV-1. OFF THE CREW DROVE INTO the lunar unknown, in what was, let me remind you, largely a 1969

General Motors product. If you've driven cars of that vintage, you know that they had as much in common with the wheezy horseless carriages of the last century's turn as they did with the computer-controlled, airbag-equipped, wonderfully reliable appliances we drive today. The rover was not a car, in a great many senses—as Sonny Morea insists, it was a spacecraft. Still, Dave Scott and Jim Irwin took off in a 1969 General Motors product *on the moon*, intending to drive it beyond sight of their only way home.

They were up for it. Scott, the mission's thirty-nine-year-old commander, was an air force colonel, test pilot, and veteran of two previous spaceflights. He had been aboard Gemini 8 in March 1966 when a malfunctioning thruster sent the two-man capsule into a potentially lethal spin; it was overcome by his crewmate, Neil Armstrong, just before the two blacked out. Three years later, Scott served as command module pilot of Apollo 9, which first took the lunar module into space and assessed its readiness for missions like this one. He was not, in other words, daunted by a little drive, even if it was in the deadliest environment a human driver had ever experienced. The Texas-born West Point graduate was a married father of two.

Irwin, Apollo 15's lunar module pilot, was forty-one and a lieutenant colonel in the air force. Born in Pittsburgh, he had talked about going to moon since childhood. His chosen route took him through the U.S. Naval Academy in Annapolis, Maryland, and into the cockpit of numerous jets as a test pilot. In a 1961 plane crash, Irwin had broken both legs, nearly losing one of them, yet willed himself back into shape to qualify as an astronaut. He also had four kids. Clearly, he didn't scare easy.

Of course, there was a backup plan: if the rover broke down, they would ditch it and walk home to the lunar module. How far they could walk was limited by the stores of air and cooling water in their backpacks. At most, they could cover about six miles. So, while the rover had enough juice in its batteries to cover forty or more miles

(Boeing claimed fifty-seven), wherever they drove had to fall within that six-mile radius.

Because their air and water would dwindle over the course of their workday, they'd start each drive by heading to their farthest planned destination, then work their way back toward the comparative safety of the lunar module. On this foray, Scott and Irwin were scheduled to drive about three straight-line miles to a mountainside, stopping at a nearby crater on the way. They would sample rocks at both places and make a couple of additional stops on the return journey.

That first drive invited sensory overload. Mount Hadley, deep in shadow, dominated the view to the north. The mountain erupted from the plain without a foothill; at fifteen thousand feet from base to peak, it outreached the Himalayas and dwarfed the Rockies. To the south rose Mount Hadley Delta—their destination today—about three thousand feet shorter but still lofty enough to rank among the great mountains of Earth. Off to the east between those massifs stretched the Swann Range, lower peaks that Scott and Irwin had informally named after Gordon Swann of the Geological Survey, who was overseeing the mission's science. To the west, and dead ahead, lay the canyon, the Hadley Rille.

Within minutes, they came to a rapid series of insights. One: What had appeared smooth plain was not, not at all. It was rippled with rises and falls, littered with small rocks, and everywhere pocked with craters. "Well," Scott said a couple of minutes into their drive, "I can see I'm going to have to keep my eye on the road."

Two: Even with their navigation system and maps, reading their surroundings was difficult. Landmarks were surprisingly elusive. Like the Apollo 14 astronauts, they found that without trees, clouds, or other visual touchstones, it was hard to pin down distances. Hadley Delta, looming high overhead, looked close enough to touch.

Three: Their heading relative to the sun played a key role in their ability to see where they were going. When they looked "down sun," or with the sun directly behind them, the details of the topography disappeared—shadows were hidden by the objects that cast them. On top of that, lunar regolith is so highly reflective that they were all but blind to the presence of humps, small craters, even enormous holes and dropoffs. The rover exacerbated the challenges of down-sun travel, as small craters and rocks sprang into view with the machine almost upon them. Scott's yells of "Whoa! Hang on!" punctuated their radio chatter.

And four: They wouldn't have lasted a minute without their seat belts. The rover's wheels smacked into obstacles constantly. The suspension absorbed the hard jolts, but even so, at least one wheel was often off the ground. The car rose and fell like a bucking bronco, as Irwin put it, and fishtailed like a speedboat when Scott attempted high-speed turns. And by high speed, I mean six miles per hour.

Less than a mile from the lunar module, they came to the edge of the Hadley Rille. It yawned wide before them, its sides riding deceptively gentle slopes to a floor that Irwin guessed to be about 650 feet wide. The astronauts had trained along the Rio Grande Gorge to get a foretaste of working on its rim. But the rille, the remnant of an ancient lava flow, dwarfed its New Mexico analog. Scott noted that while the canyon's far side was littered with boulders, its near side was smooth. "It almost looks like we could drive down on this side, doesn't it?" he asked Irwin.

"Stand by on that, Dave," a nervous Allen told him. The astronauts laughed. "I'm sure we could drive down," Irwin replied. "I don't think we could drive back out."

As they traveled, Houston often asked for their distance and bearing from the lunar module and offered estimates as to how much farther they had to go to their next waypoint. The rover's navigation system thus enabled them to estimate where they were at any moment. Minus any compass to guide them, Scott had gone through a simple process back at base camp to fix their starting point for the day's travels. Mission Control already knew the exact position of the sun over the Hadley Plain and had a pretty good bead on where the lander had set down. The rover's instrument display included a hinged, triangular "sun shadow device," essentially a sundial that, once unfolded, threw a needle-thin, precise shadow onto a graph etched into the panel's face. Scott parked the rover so that it faced down-sun, and this reading gave Houston the information it needed to "zero in" the LRV's navigation.

Or most of it: The sun shadow device's accuracy depended on the rover being perfectly level. A slight tilt to the right or left, or nose- or tail-high, threw off the angle of the sun to the sundial. So, attached to the console's left side was a two-way gauge that displayed the rover's angles of roll and pitch. Scott relayed those measures to Houston, which used them to fine-tune the heading to their first destination.

Twenty-six minutes into the excursion, Scott and Irwin stopped at the foot of Hadley Delta and at the edge of Elbow Crater, so named because the rille made a sharp bend beside it. Scott found Earth in the high-gain antenna's sighting device and started TV transmission. In Houston, engineer Ed Fendell took control of the camera, an improved design from RCA that boasted far finer and brighter resolution than those used on earlier landings. Using push buttons, Fendell could pan the camera left or right, move it up or down, and zoom in or out. It took three seconds for his commands to reach the rover.

While Fendell panned the camera in a slow arc, the astronauts collected rock samples, describing each find to Allen while placing it in an individual, numbered bag. They then sealed each bag and stashed it in a larger collection sack. They photographed every rock before they picked it up and then the empty place where it had lain. They also shot multiple images that would be knitted into long panoramas. And this was a brief stop: within fifteen minutes, Allen began urging them on to Geology Station 2. "I wish we could just sit down and play with the rocks for a while," Scott lamented. "Look at these things—they're shiny, and sparkly!"

"Come on, Dave," Irwin told him. "There'll be lots of them."

Dave Scott with the rover at the lip of the Hadley Rille, in a picture that illustrates just how deceptive distance and size are without visual yardsticks: Does that canyon look almost a thousand feet deep to you? (NASA)

Scott lashed Irwin into his seat—throughout the mission, that seat belt would require their combined effort to fasten—and headed for Hadley Delta. Somewhere up its side, they hoped to find a fresh crater rimmed with blocky ejecta and, in that rubble, lunar bedrock. Geologists believed these mountains had been heaved into being in the titanic meteor strike that created the Sea of Rains, one of the moon's largest maria.

The rover's nose began to tilt upward, their speed dropping as the ground steepened. Off to their right, Irwin spotted a block of stone sitting on the mountainside. He pointed it out, and Scott liked what he saw. The block measured a meter tall by a half meter thick. Its downhill side had a fillet, a drift of regolith piled against its base. The uphill side did not. Because dust runs downhill, like most everything, this reversed the expected order of things and suggested that the boulder hadn't been there long. "It looks fairly recent, doesn't it, Dave?" Irwin asked.

"Yeah, it sure does," Scott said. "It sure does."

"And it probably *is* fresh," Joe Allen chimed in from Houston. "Probably not older than three and a half billion years."

Scott was moved by the thought. "Can you imagine that, Joe?" he asked. "Here sits this rock, and it's been here since before creatures roamed the sea on our little Earth."

Fendell panned the TV camera, and the canyon swung into view. Allen, a nuclear physicist by training, told the astronauts: "We have a view of the rille that is absolutely unearthly." They were too busy to razz him about it. After examining the block, they took core samples, gathered rocks, and shot pictures. Between tasks, Scott scanned the plain below for any sign of the lunar module. He found none.

Forty-five minutes into the stop, Allen told them to wrap up their work. Back in the rover, they faced their first long downhill run, which Scott took at an angle. He kept their speed down, mindful that any sudden move would bury the front wheels in the soft regolith and spin them out of control. Sure enough, when he steered hard to avoid a small crater, the rover's back end instantly swung around, and when it stopped, they were facing uphill. They laughed giddily. "You just—you can't go fast downhill in this thing," Scott explained to Allen, "because if you try and turn with the front wheels locked up like that . . . around you go. And we just did a one-eighty."

Down near the mountain's base, the ground was littered with rocks and boulders. Among them, the familiar caught Scott's eye. "There are some rover tracks!" he cried. "How about that?"

Irwin responded with the best line of the mission: "Somebody else has been here."

They were close to Hadley Base when Scott spied a chunk of black basalt sitting by itself on the light gray plain, so seemingly out of place that it brought to mind the rocks the Geological Survey had seeded among the craters at Cinder Lake. He wanted that rock—but knew that if he asked for permission to stop, Allen would turn him down: Mission Control wanted them back to set up the remote ALSEP array, which would take hours. Instead, he pretended he was having trouble with his seat belt. "Okay, we're stopping," Irwin said—then, catching on to his partner's ruse, launched into a monologue about the small craters and rocks around them. Scott was back quickly. The rock would become known as the "seat belt basalt."

55

THAT FIRST DRIVE CULMINATED A GENERATION'S THOUGHT AND LA-bor to put wheels on the moon. It had been nineteen years since von Braun's stories in *Collier's*, and fourteen since Greg Bekker's first exploratory soil bin tests with "lunar" ground pumice. Hundreds of people had devoted themselves to the dream since. Now it seemed worth the trouble: LRV-1's first journey had been flawless, save for the front steering.

The second drive erased even that small qualification. When Scott activated the rover the next day, he discovered that, for reasons neither he nor Houston could fathom, the front steering worked. "You know what I bet you did last night, Joe?" he asked Allen. "You let some of those Marshall guys come up here and fix it, didn't you?"

He and Irwin set off a short while later. Their destination was again Hadley Delta: they'd drive straight to the mountain and climb its flank higher than they had the day before, and there sample multiple sites for older rock than they were likely to find on the plain—old enough, perhaps, to bear witness to the moon's creation. Geologists theorized that the ancient lunar crust was composed largely of crystalline rock called anorthosite. That was their quarry.

Four-wheel steering differed sharply from the rear-only of the previous day—it was "a new task," as Scott put it. "It's really responsive now. I guess I got pretty used to quiet steering, and this thing really *turns*."

"Roger, Dave," Allen said. "We don't want it to be too easy for you."

They crossed the corduroy plain, Irwin providing a running description of the surrounding country. Within minutes, Scott interrupted him—the steering was "a little *too* responsive." They stopped while he turned off the rear. They resumed for only seconds before he stopped again. The back wheels had not locked pointed straight ahead, as they were designed to do; he sensed they were loose, drifting. He turned the steering back on.

They set out again, roller-coastering in and out of old and eroded craters, accelerating to six miles per hour. Whenever Scott's gaze strayed from the ground directly ahead of them, he almost always regretted it; the ground was perforated with younger, sharper-edged, deeper holes, and they slammed into several. Rocks appeared with increasing frequency and grew in size—six inches to a foot across, most of them, though some measured two feet. The rover had fourteen inches of ground clearance. Scott cut a snaking route to avoid the biggest.

Ahead, Hadley Delta's smooth face seemed to reach an impossible distance into space. They could see all the way to its crown, twelve thousand feet up, and the closer they drew to its base, the more inconceivably massive it appeared. "Boy, that's a big mountain when you're down here looking up, isn't it?" Scott said. "That's as big a mountain as I ever looked up." Forty-two minutes and more than three miles from the lunar module, and 240 feet above it, they stopped on mountainside littered with rocks.

"By golly, Joe, this rover is remarkable," Scott reported as they struggled to find their feet on an eleven-degree incline. This time they could see the lunar module, a distant blip of silver and orange in a field of gray. "Boy, what a view, huh?" Irwin sighed.

They gathered samples. "You know," Scott said, pausing to look back across the Marsh of Decay, "we're a long way from the LM."

Dave Scott took this photo to document the green boulder, but captured Jim Irwin braced against the rover to keep it from sliding down the mountain. (NASA)

Next leg: uphill to a squat, solitary boulder they could see from far below. When they stopped, 312 feet up the mountain and just above the rock, the ground underfoot sloped even more severely. Irwin regarded the downhill route to the boulder. In addition to being steep—eighteen degrees, by postmission reckoning—the soil here was loose. "Gonna be a bear to get back up here, you know."

Scott suggested that Irwin let him attempt the descent first, while Allen interjected a nervous plea that they not stray too far from the rover. Taking a moment to consider his options, Scott changed his mind: "Let me drive the rover down there," he said. "Could you watch me as I back up here?"

Irwin shuffled away from the machine. "Sure can."

"Got your eye on me?"

"I have my eye on you," Irwin confirmed.

Scott backed the rover sideways across the mountain face—a moment at which he could have used the eliminated rearview mirror—then pointed it downhill. "How am I doing?"

"You're doing okay," Irwin told him.

"Roger, Jim and Dave," a tense-sounding Allen said. "Proceed very carefully now, please." Scott eased the rover forward. It crawled down the slope, tires sliding. When he brought it to a stop below the boulder, it barely clung to the uneven mountainside and its left rear tire hung several inches off the ground. It shifted as he climbed off, prompting a worry: What if it took off down the mountain? "Tell you what, Jim," he said. "We'd better abandon this one."

"Afraid we might lose the rover?"

"Here, you come down and get on," Scott said.

Irwin, who had paused beside the boulder, offered an alternative. "Let me hold that rover, and you come up and look at this," he said. "Because this rock has got green in it, a light green color."

"Okay," Scott said, "I'll just stand here until you're through, and then I'll go up and take a look at it." Scott had not aimed the TV antenna, so Mission Control only heard the exchange. Irwin's reference to losing the rover did not calm the room. "Roger," Allen said, reminding the astronauts that they had an audience. "We're copying all of it." Finished with the rock, Irwin joined his partner. Scott had him stand on the rover's downhill side, braced to check its movement. Scott didn't take long. When he climbed back aboard, Irwin's seat belt remained unfastened. "I'll just hang on, Dave," he said—and it was a measure of the precariousness of their perch that Scott went along with the idea. They slipped and slid and fishtailed their way downhill, Irwin holding on with both hands. Allen suggested he turn on their sixteen-millimeter movie camera for footage of the rover under way. "No chance right now," Irwin told him.

When they climbed off the machine at Spur Crater, their second excursion had already lasted as long as their first. They sampled the ejecta along the crater's rim, finding more greenish stone and soil, too. It stood out in their color-leached surroundings, and they wondered whether their eyes were tricking them. They collected a couple of samples—the rock *was* green, by the way, thanks to the presence of olivine glass, a remnant of the moon's distant past—and scanned the crater for their next sample. That's when Irwin noticed a rock pedestal jutting from the surface with "a little white corner" on its upper edge. Once he had taken pictures, they moved in on it. Scott pulled off the white nugget with a pair of tongs. It was about four inches long, half as wide, and weighed nine and a half ounces. He turned it over in his glove. "Oh, man!" Irwin gasped.

"Oh, boy!" Scott said. "Look at that."

"Look at the glint!" Irwin said.

This was no ordinary moon rock, if such a thing exists. It was a chunk of bright-white, nearly pristine anorthosite—a time capsule more than four billion years old. "Guess what we just found?" Scott said as Irwin laughed. "Guess what we just found? I think we found what we came for."

They continued working the site and, in the course of raking the soil, collected three smaller pieces of anorthosite. But the big one, which the press dubbed the "genesis rock," earned worldwide attention as the oldest piece of the moon yet sampled and remains one of the most important scientific finds of the Apollo program. In Houston, Apollo 15's lead flight director, Gerald D. "Gerry" Griffin, no doubt had the discovery in mind when he called the second EVA "the greatest day of scientific exploration, certainly, that we in the space program have ever seen—and possibly of all time."

And it was, to that point: the moment that Scott and Irwin picked up the genesis rock, recognized it for what it was, and told Houston what they'd found was the consummation of all the missions that had come before, dating back ten years to Alan Shepard's humble fifteen-minute flight in the first manned Mercury capsule. Until now, most missions had been built around testing equipment and sorting out procedures. Apollo 15's moonwalkers were conducting real science, far from the safety of their lunar module. They truly were exploring another world.

A half hour after the discovery, the astronauts climbed back into the rover. They had been in the field for three hours and were nearly three miles from the lander. If the rover conked out, they would be bumping up against their supplies of air and water on the walk back. Houston wanted them to get a move on. They drove downhill, found their outbound tracks, and followed them back. After a brief stop at a crater along the way, they beelined for the lunar module. "Gee, it's nice to sit down, isn't it?" Scott asked as they cruised.

"Oh, it is," Irwin agreed.

Scott laughed. "It's a good deal. You hop off and work like mad for ten minutes—and hop on, sit down, and take a break."

56

APOLLO 15 DID NOT GO OFF WITHOUT A HITCH. THE SIXTEEN-millimeter movie camera broke early in the mission, so the crew came home with very little footage of the rover on the move. The failure also nixed a planned "lunar grand prix," in which Irwin would film Scott putting the rover through a prescribed series of maneuvers—something that Sonny Morea, his Marshall Center colleagues, and the GM team were keen to see. Irwin's still camera also quit working late in their travels, which complicated the crew's documentation of samples.

But the most persistent annoyance was a handheld electric drill they used at the ALSEP site to excavate deep core samples and holes for a heat-flow experiment. The drill's flawed design cost the men time and effort when they returned from their first two drives. As they champed to get started on their third, Houston had them wrestle with the thing for another half hour. Frustrated, Scott asked their capcom, "How many hours do you want to spend on this drill, Joe?"

Then, two minutes later: "Joe, I haven't heard you say yet you really want this that bad." Later still: "Hey, Joe, you never did tell me that drill was that important. Just tell me that it's that important, and then I'll feel a lot better."

Allen finally responded: "It's that important, Dave." And indeed, they extracted a core sample offering a detailed record of geologic events spanning eons. But lunar stays are necessarily zero-sum games, and the delay, along with the promise of struggling further with the drill at the EVA's end, cut into the time available for the third drive. When they finally got going, their last excursion was sure to be truncated.

Scott and Irwin headed west, straight toward the Hadley Rille, moving along at a brisk five miles an hour, and discovered that, while the surface was smooth and relatively free of rocks, it rose and fell like Saharan dunes. The lunar module disappeared behind the swells. "You really could get lost here," Irwin observed.

The rolling surface hid a chain of huge, subdued craters, their sides gentle but their floors one hundred feet down, sometimes more. They drove around a couple and straight through another. Progress was slow. "I thought we'd whip right over to the rille," Irwin said. "I didn't think we'd have this type of terrain."

A mile from the lander, they stopped at a small, fresh crater rimmed with what Scott called a "nice ejecta blanket." When they aimed the high-gain antenna, they saw the TV camera was drooping, something they'd noticed it starting to do the day before. They leveled it manually so that Ed Fendell could follow them around the crater's edge. A second short drive brought them to the edge of the rille. They could see miles up the gorge in both directions. Arizona's gigantic Meteor Crater would fit inside it like a cereal bowl in a kitchen sink. Indeed, the rille's breadth was so vast that it clouded one's sense of its depth. Away to the south, the canyon made its abrupt bend at the foot of Hadley Delta, the sheer mass of which further scrambled one's perception of scale.

They raked for samples and collected rocks. At one point, Scott descended the rille's side on foot as the TV camera followed him. The slope was gentle and the footing good, but on the screens at Mission Control, he seemed at the edge of a precipice, prompting Allen to fret. "How far back from the lip of the rille do you think you're probably standing?" the capcom asked.

"Can't tell," Scott breezily replied. "I can't see the lip of the rille."

Soon Allen was urging them on to a point a quarter-mile north to photograph the canyon's far wall. It was easier driving, much smoother than their trip from the lander, and they arrived a little over two minutes later. "All we need is photography from this stop," Allen told them, "and we're looking toward arriving back at the LM in about forty-five minutes." The original plan had been to continue north from there, across the plain to a cluster of hills, possibly volcanic, known as the North Complex. Not anymore. While Scott took telephoto images of the rille, Irwin tried to get in some last-second geology, describing a boulder in some detail. Allen cut him short. Their time was up. They took a straight-line route to base, using the navigation system. It was right on the money.

Back at the landing site, the core sample required their attention. Afterward, Scott and Irwin organized the rocks they'd collected and lugged them up the ladder and into the lunar module by the bagful. They unloaded gear extraneous to the journey home—with their new cargo of geologic samples, every pound counted. Allen asked for a photo of the deployment gear saddle that had gotten hung up. Scott snapped one.

The TV camera struggled to follow them as they conducted their chores, and the astronauts paused to help it along. At one point, Allen asked them to realign the antenna. With that task completed, Scott suggested that Fendell follow him to the equipment bay on the side of the lander. Irwin had to point the camera manually.

"Okay," Scott said to the audience at home. "To show that our good postal service has deliveries any place in the universe, I have the pleasant task of canceling, here on the moon, the first stamp of a new issue dedicated to commemorate United States achievements in space." He produced an envelope adorned with stamps depicting the lunar module and two astronauts riding the rover, and canceled them with a rubber stamp. "By golly, it even works in a vacuum," he said, eyeing the envelope, then added: "But not too well." Next, he tested an age-old scientific theory for the camera, simultaneously dropping a three-pound hammer and a falcon feather weighing just over a gram. They hit the regolith at the same time. "How about that!" Scott said as applause erupted at Mission Control. "Which proves that Mr. Galileo was correct in his findings."

Those moments of theater finished, Scott climbed onto the rover and drove it over the hummocky surface to a slight rise about 540 feet away. There he turned it around so that its nose faced the lander, engaged the parking brake, climbed off, and dusted the electronics. Off camera, in a small crater twenty feet from the rover, he created a memorial honoring astronauts and cosmonauts who had died on the job or while preparing for space—a plaque bearing fourteen names, along with a small, stylized aluminum figurine, *Fallen Astronaut*, by Belgian sculptor Paul Van Hoeydonck. He balanced a red-bound Bible on the rover's armrest, leaning it against the hand controller. Then he aligned the high-gain antenna and lifted the drooping TV camera until it was level. LRV-1 had one last duty to perform: on the first three moon landings, the astronauts had blasted back into space unseen. This time the rover would provide live TV coverage of *Falcon*'s departure.

It happened fast. Viewers saw the engine glow under the ascent stage, then a spray of sparks as the cabin shot up and out of the picture. Ed Fendell might have followed it, but the clutch for the camera's tilting function had given up the ghost.

Four days later, as Scott and Irwin neared Earth with their command module pilot, Al Worden, Houston called with news from the abandoned landing site. Mission Control had reactivated the rover's TV camera, Joe Allen said. It worked "beautifully" for about thirteen minutes—Fendell panned the camera and zoomed in on the mountains. Then, without warning, the signal quit, and the picture vanished. "We most likely popped a circuit breaker or something like that," Allen said.

"Would you like us to go back up and check it for you?" Scott asked.

"Knew you were going to ask," Allen told him.

In the weeks after the mission, the Marshall Center obsessed over the rover's minor failures—in particular, the front steering glitch during the first EVA. The Astrionics Lab was on the case even as Apollo 15 left the moon, theorizing that the snag was most likely due to an open circuit or a mechanical hang-up that worked itself out during the first drive. By late in the month, the labs were considering a third culprit: the potentiometer in the hand controller that Pavlics had suspected. They never pinned it down. Houston, not quite so obsessive, blamed "MEF," or mysterious evil forces.

Just about everyone else marveled at the little machine's performance. LRV-1 had proven safe and reliable. Its wheels maintained traction even in the soft, steep soil of Hadley Delta. The motors and harmonic drives worked flawlessly and ran at far cooler temperatures than expected. The battery temperatures, while higher than predicted, were never too high. The rover's travels used about half the power that mission planners had forecast, and its suspension toughed its way over some very rugged terrain. "We had almost no right to expect the rover to work so well," Apollo program boss Rocco Petrone told reporters. "It made the mission."

Moreover, the machine had upended all assumptions about how much science could be accomplished in a single visit to the moon. Scott and Irwin spent eighteen hours and thirty-five minutes outside—impossible without the rover. They drove for just over three hours, during which they covered 17.25 miles. At one point, they ranged 3.2 miles from their lander. They brought back 170.4 pounds of samples.

Millions of Americans watched the rover in action, and the media covered *Falcon*'s three-day stay with gusto. "In brief, the LRV performed as planned," Morea wrote rather modestly in his weekly notes, "and none of the problems with the LRV can be classified as any worse than a deviation." The rover so impressed the rest of NASA that Lee Scherer, director of the Apollo Lunar Exploration Office at headquarters, promised it would be trusted to assume an even bigger role in the missions to come. "We are going to plan longer traverses now," he said, "than we would have planned several weeks ago."

In Santa Barbara, the GM team celebrated. On the evening of August 7, 1971, Romano, Pavlics, and another GM engineer, Joe Finelli, cohosted a splashdown party at Romano's place. Hundreds of guests jammed the house and backyard. "It was crazy," Romano's son Tim told me. "There were cars lined up and down the street. People were everywhere—neighbors and people from GM. It must have been everyone they knew."

57

THE SECOND ROVER, DELIVERED TO THE KENNEDY SPACE CENTER that September, underwent a few small modifications in the wake of Apollo 15. The most obvious changes involved the seat belts, the one serious beef that Dave Scott and Jim Irwin had voiced. Scott suggested replacing them with "a bar-type affair like in the carnivals—a little kiddie bar," arguing that it "would save considerable time and effort" on the moon.

NASA wasn't much interested in a kiddie bar but did take the comments to heart. Apollo 16's Charlie Duke tried out a replacement belt on the Vomit Comet, suggested a few refinements, and flew with it again just to be sure. He and his mission commander, John Young, signed off on the new design—adjustable and far easier to clasp—after yet another flight at simulated one-sixth gravity in early November. The rover team hurried to build and install the belts. Technical issues postponed the launch from its original date—March 17—to Sunday, April 16, 1972. Early that afternoon, Apollo 16 lifted off from the Cape.

Later in the day, the rocket left orbit and shot for the moon. The command service module *Casper* loosed itself from the third stage, performed an about-face, and docked with the stored lunar module, *Orion*. Together, on they flew. A guidance system problem threatened to end the mission early, but the crew worked around it. Four days after launch, as *Orion* prepared to descend to the lunar surface, *Casper*'s monitors signaled an engine glitch. That almost scrubbed the landing, but after the two craft hung side by side in orbit for nearly six hours, Houston gave its okay to proceed. The lunar module descended to the rugged Descartes Highlands with LRV-2 pinned to its side.

Orion came to rest on a mash of mountains, valleys, and undulating plains pocked with craters and littered with boulders. It was the southernmost Apollo landing site and, far removed from the maria on and around which previous visits had occurred, the highest in altitude. The optics from the lander offered a lot of promise. To the south rose Stone Mountain, round shouldered and 1,800 feet tall. West of the mountain and about three miles from camp gaped the steep-sided South Ray Crater, rimmed in bright white ejecta, and an equal distance to the north, another major landmark, North Ray Crater. The region marked the seam between two geological formations—the Cayley and Descartes—that geologists believed to be volcanic. They were eager for rocks to prove it.

As the crew originally chosen for the rover's first outing, the men aboard *Orion* had spent a lot of time in Huntsville and Santa Barbara, and were well known to GM's rover team. "They followed very closely the vehicle's design stage, so they were very close to us," Frank Pavlics recalled. "They were the most fun to work with—good old Southern boys."

Indeed they were. At times, their banter on the moon seems lifted from *Smokey and the Bandit*. John Young, forty-one, was a navy captain and former test pilot making his fourth spaceflight. He flew on the first crewed Gemini mission in 1965, and commanded Gemini 10 the following year. In May 1969 he piloted the command module on Apollo 10, the dress rehearsal for the first moon landing two months later. Raised in Florida, Young had a reputation as an astronaut's astronaut—smart, funny, and unflappable.

Charlie Duke, the mission's lunar module pilot, spent most of his boyhood in South Carolina and, at thirty-six, was the youngest person to walk on the moon. An Eagle Scout, Naval Academy graduate, and air force test pilot, he was selected in NASA's fifth group of astronauts, along with Jim Irwin. This was Duke's only spaceflight, but his was among the Apollo program's familiar faces: he'd been capcom when Apollo 11 made the first moon landing. When Neil Armstrong announced, "Houston, Tranquility Base here. The *Eagle* has landed," it was the drawling, garrulous Duke who answered: "Roger, Tranquility. We copy you on the ground. You got a bunch of guys about to turn blue. We're breathing again. Thanks a lot."

Duke had also earned some notoriety when he was exposed to German measles while a backup crewman on Apollo 13. The only member of the prime crew not immune to the disease was Ken Mattingly, the command module pilot. Though he didn't get sick, NASA yanked Mattingly from the mission three days before liftoff, replacing him with his backup, Jack Swigert. Now, two years later, Mattingly was orbiting above Young and Duke in *Casper*.

Landing hours late, the pair caught a few hours of sleep before first venturing from *Orion*. Then, on Friday, April 21, Young descended the module's ladder and stepped onto the moon. "There you are," he said. "Mysterious and unknown Descartes Highland Plains, Apollo 16 is gonna change your image." He added a reference to his own long career in space: "I'm sure glad they got ol' Brer Rabbit, here, back in the briar patch where he belongs." Duke, impatient to join him, gave Houston a one-sentence status report on the lander, then yelled, "Here I come, babe!" His first words on the surface were: "Hot dog! Is this great!"

They unloaded the rover first thing. "Believe it or not, this is like in the training building," Duke said, descending the ladder after pulling the D-ring. "The only thing we don't have, Tony, is the linoleum on the floors." He was addressing their capcom, Tony England, an MIT doctoral candidate and a scientist-astronaut since 1967. Once Young and Duke freed the machine from the lunar module, they picked it up, turned it around, unfolded its seats and console, and attached its toeholds. Young climbed in to pull it away from the lander. "Ah, this is going to be some kind of different ride," he said.

"The rover is running, Houston," Duke reported. Then, watching Young bounce slowly over the regolith, he told him: "Okay, your rear steering's off."

"Huh?"

"You don't have any rear steering," Duke said.

After fiddling with switches for a minute, Young decided to leave the fix until later; they were on a tight schedule to set up the ALSEP instruments before they drove. Besides, he wasn't worried: Young claimed later that they would have used the rover with no steering at all. "We planned to go in a straight line as far as we could, get out, pick the thing up, and point it in the right direction and keep right on going," he said. "It would have been a lot better than walking."

They loaded the rover with the ALSEP gear. Before driving away from the lander, Young tried the rear steering again. This time it worked. For the second mission in a row, the LRV had lost half its steering for reasons unclear, only to have it return just as mysteriously. "Maybe it just needs to sit around and heat up," Young suggested.

Arranging the ALSEP took a little more than three hours, during which Young broke one of the experiments by accidentally tangling himself in a cable. Finally, they climbed into the rover. Their first drive would be short, taking them just a mile west of the lander—past

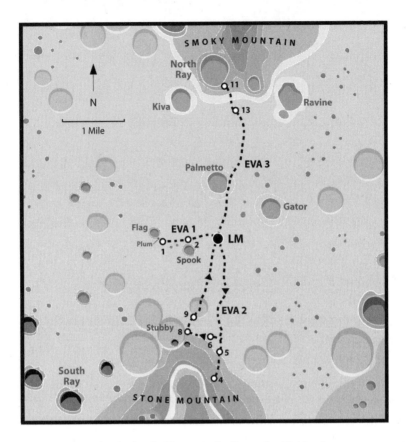

The three drives by the Apollo 16 crew on the Descartes Highlands.
(KEVIN SWIFT)

the wide but shallow Spook and Buster Craters on their way to the eight-hundred-foot Flag Crater. There they would collect samples and visit a smaller crater, Plum, punched into its rim.

Driving was a challenge. Like moonwalkers before them, Young and Duke found distance impossible to judge. They headed down-sun, all but blinded to hazards ahead. Rocks and small craters—three feet across, five feet, ten—were everywhere. "There's a big crater," Duke warned. "There's about a ten-meter off to your left there, John."

"I see it," Young said.

"A deepy one over here," Duke said, looking to the right.

"Charlie, you hit my arm."

"Excuse me."

"I'll end *up* in that big crater." In their pressurized suits, the astronauts were wider than normal, and their shoulders touched as they rode. "It was *very* sensitive steering," Duke told me. "If anything, the steering was maybe a little too sensitive—just a little jiggle of the handle would cause the front wheels to move around. We'd be under way, and I'd turn in my seat to get a side picture, and I'd bump John's elbow, and that would . . . change his steering, and then he'd start fishtailing as he tried to get it back under control. Depends on how fast you were going, but it had a lot of fishtailing. It was squirrely, like driving on snow and ice."

Young kept the rover's speed to about three miles per hour, struggling to see through the glare and mindful that geologists believed the area to be crisscrossed with scarps, or sudden drop-offs. Duke, however, was ready to let it rip. "Hey, man," he said, "we could just *go*, babe. I'm really cinched into this moose."

"Yeah, well, but I don't know, with these holes, if we ought to do that or not."

England asked whether the rocks they saw back at the ALSEP site had been breccias: mishmashes of various rock types, fused by the shock of meteor impacts. Nonvolcanic.

"Charlie," Young growled, "you hit my arm."

"I'm not sure, Tony," Duke answered. "I think they were breccias, but they were really dust covered, so I couldn't tell you, really."

"Okay, understand. And have you seen any rocks that you're certain *aren't* breccias?"

"Quit hitting my arm!" Young complained.

"Negative," Duke told England. "I haven't seen any that I'm convinced is not a breccia."

Their first stab at lunar navigation proved tumultuous. They had almost passed Spook before they recognized it as Spook. They rolled past Buster Crater, with England warning that "there should be a scarp around there someplace." There wasn't, and a minute later, they weren't so sure they'd passed Buster—the next crater they came to matched its description, too. Meanwhile, Duke had to turn down the cooling garment under his suit. He had set it on the middle "intermediate" setting while working on the ALSEP. Now, with the lower metabolic rate that came courtesy of the rover, he was freezing.

Young thought he saw Flag Crater up ahead. "It'll do," he said. "Let's not waste any more time thinking about it." He pulled up next to the crater and parked.

"Okay, Tony," Duke radioed. "We're going to call this Flag." Once out of the rover, though, he had misgivings—it looked to him like this might be yet another crater, Halfway, that they knew to be somewhere close. "This might be Halfway, this one right here," he told Young.

"This one here?"

"Yeah," Duke said. "This one right here."

England cut in. "Our measurements say that you should be pretty near Halfway."

They climbed back aboard, and Young eased the rover forward. "I feel real faith in this thing," Duke said. "Open her up a little bit."

"I can't see where I'm *going*, Charlie." A couple of slow-rolling minutes later, they approached another depression. "That's it," Young said.

"Okay, Tony," Duke reported, "I think we finally found Spook, here. Or Flag, rather." He read their bearing and range to the lunar module off the rover's instrument panel. England confirmed that Flag Crater was right where it was supposed to be.

Once parked, Young aimed the high-gain antenna. On Mission Control's screens, there appeared a scene of strangely plastic forms, devoid of hard edges. Flag, which the crew had named for Flagstaff, where they'd conducted so much of their training, was so deep that the astronauts couldn't see its floor, but its rim was soft and flowing. Plum, just 130 feet across, was likewise rounded and smooth, though its sides were treacherously steep.

Rock hammer in hand, John Young closes in on a prospective sample near the lip of small but steep-sided Plum Crater. (NASA)

"We gave a wide berth to the rim of that crater, because if you fell in, you wouldn't be able to get back out," Duke recalled. "You get down in a crater like that, and it sloughs. You can't get traction. The angle of repose is such that if you try walking out, you keep sliding backward. We had to be very careful, because we had no rescue."

It's worth taking a moment to chew on that. No rescue. A slight misstep, a minor mishap on Earth, was deadly in the Descartes Highlands. Watching the rover's TV feed of the astronauts bunny hopping and skipping across the surface made the adventure look more playful than dangerous, and watching that footage today, fortified by the knowledge that it all turned out well, can jade one to the hazards involved. The fact was, staying alive required vigilance.

The geologists in Mission Control's back room detected something interesting on their screen and passed word to England. "Over to the right on the edge of Plum there, looks like there might be a rock with some phenocrysts," or large crystals, the capcom told the crew. "If you're over that way, you might look around and see if you see something like that."

In the meantime, they raked lunar soil and dumped it into sample bags, collected rocks, and took panoramic photos. On their way back to the rover, England again brought up the rock. "We'd like you to pick it up as a grab sample," he said. Duke pointed to a large block of stone sitting on Plum's lip. "This one right here?"

"That's it," the capcom said.

Duke, somewhat incredulous, pointed with a long-handled scoop. "*This* one, right here?"

"That's it," England said again. "You got it. Right there."

"Okay," Duke said. "That's a—"

"That's a football-size rock," Young interrupted. "Are you sure you want a rock that big, Houston?"

"Yeah," England said. "Let's go ahead and get it."

Duke dropped to his knees and tried to roll it onto his right thigh. When it fell, he sandwiched the rock between his hand and leg and struggled to his feet. A quick hop sent the rock floating upward, and he grabbed it with both hands. "If I fall into Plum Crater getting this rock," he told England, "Muehlberger has had it." University of Texas professor Bill Muehlberger was the mission's top geologist. He also inspired the rock's name—"Big Muley." It was the largest single piece of moon brought back by any Apollo crew. "Okay, I've got it," Duke said. "That's twenty pounds of rock!"

Almost twenty-six, actually—and another breccia. They found no volcanic samples at their first stop. They drove on to their second, following their tracks back to Buster Crater.

The second stop's geology was as confusing as the first: nothing but breccias. "We were briefed that the Descartes region was two volcanic flows, one more viscous than the other, and [they met] in the Cayley Plain," Duke said when we spoke forty-seven years later. "We were to sample back and forth across that contact. So, when we got there, we started describing these rocks, and they weren't volcanic at all. I got the feeling that their attitude was, 'Are these dummies? We wasted six years on [training] them. They don't know what they're doing.'"

Back at the lander, Duke used their sixteen-millimeter camera to record the second attempt at a grand prix. Young's instructions: take LRV-2 through a succession of sprints and hard turns, enabling engineers back home to study the machine's driving manners. With Duke urging him on, he jammed the stick forward. One astronaut lighter than usual, the rover rose and fell like a small boat on a choppy sea as he built up speed. When it hit bumps, its front wheels hung in space for a weirdly long moment, as if in slow motion. The rover arced to the left, pitching and rolling, and motored toward the lunar module. "Is he on the ground at all?" England asked. With the rover moving, Mission Control was blind to the action.

"He's got about two wheels on the ground," Duke replied. "There's a big rooster tail out of all four wheels. And as he turns, he skids. The back end breaks loose just like on snow." Young turned the rover around and headed back. "Man, I'll tell you, Indy's never seen a driver like this," Duke sportscasted. The rover plowed into small craters, throwing dust. The suspension, working hard, did a remarkable job of smoothing out the bumps. As the car jounced toward the camera, each wheel crazing up and down, the chassis bucking, it looked like the whole thing was made of rubber. (Boeing's Gene Cowart recalled watching the footage later and being "amazed at how it rocked and rolled. I sure didn't expect it to flop around like that.") Young threw it into a turn and locked up the brakes. Dust flew.

Houston wanted more. Once again, he took off for the lander. When he was about one hundred yards out, Duke said, "Turn sharp."

Young chuckled. "I have no desire to turn sharp." Then, changing his mind, he spun the rover to the left. The rear end swung around as he threw the stick forward. Back he came, corkscrewing across the surface. "The suspension system on that thing is fantastic!" Duke said. Despite the rover's bouncing, Young seemed in complete control. As it neared Duke, the car launched out of a crater. "Man, that was all four wheels off the ground, there," Duke said. "Okay, max stop."

This time Young demurred. He brought the rover to a rolling halt. "Okay," he said. "I have a lot of confidence in the stability of this contraption."

Back at the foot of the ladder, they dusted each other off. Duke was appalled by how much black dust had collected on his space suit. "I am so dirty, I can't believe it," he said.

But at that point, as he'd soon find out, he didn't know dirty.

58

RESTED AND FED, THE MOONWALKERS OF APOLLO 16 BEGAN SATurday, April 22, 1972, with several hours of housekeeping—or "overhead," as they called it—before their second drive. They set out on a heading for Survey Ridge, which rose from the plain just shy of a mile away. The ground, cobbled with rock and perforated by craters, made for a rough ride.

Up on the ridgetop, it got worse. Young dropped their speed and steered among boulders and craters as best he could, but even so, the front wheels slammed into rocks or dropped into pits, heaving dust over the car. "Boy, you just can't believe the blocks, Houston," Young said. To Duke, he added: "Got to get out of these, Charlie."

They altered course to escape the cobbles, which were ejecta from the nearby South Ray Crater. Down the far side of Survey they rolled, bottoming out on the rims of small holes. "I'm even getting dust on my helmet," Duke reported.

Young, laughing: "This is the wildest ride I was ever on."

Their first stop waited up on the side of Stone Mountain. The climb began imperceptibly; they were already gaining altitude when Duke wondered aloud whether they were on a slope. Young didn't think so. Soon enough, all doubt vanished. The rover clawed up an increasingly steep pitch. Young took it at an angle, so the machine was rolled five to ten degrees to the right. It was a modest amount, but as open and unprotected and high off the ground as they sat, felt extreme. They also noticed that the dial on their attitude indicator—the small device on the console that registered the rover's pitch and roll—had popped out of place, leaving just the needle. As they continued to climb, it swung to its twenty-degree maximum.

They stopped on a small shelf, the only marginally level spot they saw on the mountainside, and for the first time, Duke looked back the way they'd come. They were five hundred feet above the plain, far higher than he expected. "It was spectacular," he recalled. "The most encompassing view and the most spectacular view that we had. And out there in the middle you could see a little dot of orange"—the lunar module. "It was the only color."

Young gasped as he absorbed the sight: "It's absolutely unreal." Later, he recalled that as he looked down the mountain, he had the sense "we might have bitten off at least as much as we could chew." Even on the shelf, the ground was so tilted that the rover's right rear wheel dangled in space. "That thing doesn't look like it's too stable," Duke observed, but a better parking space was nowhere to be found.

They collected samples, but again found no volcanic rocks. "They had fragments of volcanics in them, but you could tell they were pyroclastics that had been blasted out" in meteorite impacts, Duke told me. "So these rocks had melted together and fused up. We found a lot of breccias." Their geologic duties complete, the pair loaded the rover and climbed aboard. Now that they'd be heading downhill, the mountain looked even steeper. "Okay," Young told his partner. "Now, watch my arm now, okay? Don't hit my arm."

"This is going to be sporty," Duke predicted.

They rolled down the mountain, Young braking to keep their speed in check. He knew, he said later, "that if you let that rascal loose, that she'd go down that hill in a big hurry." Halfway down, they crossed their uphill tracks and turned to follow them the rest of the way. Almost a half century later, the toboggan run down Stone Mountain remained vivid to Duke. "*Steep*," he said. "You had your feet braced on your footrest, even though you were buckled in real tight."

"No wonder we broke the pitch meter," Young said as they nosed downward. He was impressed by the rover's surefootedness. "Look at this baby. I'm really getting confidence in it now," he said. "I've got the power off, and we're making ten kilometers an hour, just falling down our own tracks."

As they rolled, Houston wanted them to watch for a fresh crater. It had to be one drilled by a meteorite rather than flying ejecta from another impact—geologists wanted Stone Mountain bedrock and reckoned that only a high-velocity strike would bullet deep enough. After a few minutes of driving, Young pointed out a candidate, and Duke agreed that it looked good. They parked at such an extreme angle that, once they were on their feet, Young suggested they reposition the rover. Rather than get back in, they picked up the machine and reparked it by hand.

After working their way around the crater's rim, bagging samples as the TV camera followed, they set off for Station 6. England told them it should be "at the lowest terrace on Stone Mountain and a blocky crater, if possible." They found such a crater, stopped briefly, then continued downhill. A few minutes later, after bouncing through a boulder field and scraping bottom on at least one large rock, Young and Duke decided to detour for a look at the ancient and eroded Stubby Crater. They had to climb a ridge to see it. Halfway up, Young complained about its steepness. "Have you got full throttle on?" Duke asked him.

"Yeah."

"Boy. We're hardly moving."

Young checked his instrument display. "What it is," he told Houston, "is we've lost the rear-wheel drive. Not reading any amps on the rear wheel." They cut off the climb and turned away from the ridge. Back on the flat, Young radioed a theory about what might have happened: they had driven over a rock "that cut a wire or something in the back."

The rover had no exposed wiring. Still, it was possible they had shaken loose a connection. At the next stop, they used hammers to whack at a boulder, collected the resulting shards, pulled a core sample, and contemplated their crippled machine. They decided to check the rear steering to see whether it was out, too. Young climbed aboard while Duke watched from a few feet away. The steering worked.

"Why don't you try rear drive only?" Duke suggested next.

Young flipped the necessary switches and pushed the stick forward. "Just sitting there, doing nothing," he said. He tried switching to the other battery. "Nothin'," he muttered. He radioed Houston. "Okay, we've tried the forward steering and rear steering. . . . We have rear steering. We have forward steering. We have forward drive power, but we don't have any rear drive power on either [battery]."

Houston suggested he throw a switch on the instrument panel, surrounded by a guard to prevent inadvertent knocks, that controlled the "drive-enable" functions to both ends of the machine. Young saw at once that the switch was at the wrong setting. "Oh, that's the problem!" he said. But no: when he turned off power to the front wheels and worked the hand controller, again the rover sat. England suggested they return the rover to its default configuration. Young did so, then pushed the stick. That worked. "Okay," Duke told him. "You've got it now."

In the minutes before they solved the problem, the phone rang at Frank Pavlics's house in Santa Barbara. His ten-year-old son, Peter, answered. It was the NASA operator in Houston, trying to establish a conference call between Mission Control, the Marshall Center, and Pavlics to discuss the rover's troubles. The operator asked for his father. In the background, Peter could hear the radio chatter between the astronauts and Tony England. He ran into the backyard, shouting, "Dad! They are calling you from the moon!"

At the next stop, Young and Duke chipped pieces from boulders and hauled in a pile of rocks, all of scientific interest. But the stop would also prove important to our story on account of an accident. As he was moving around the rover, Young passed too close to the right rear wheel, and either his suit or a hammer jutting from his shin pocket snagged the fender. With a spray of dust, its sliding extension snapped off. "There goes the fender," he said.

On the drive to Station 9, neither astronaut mentioned the effects of the missing piece, but they were pronounced. Without that few inches of fiberglass, the right rear wheel flung a steady arc of dirt over the rover and its passengers. After they stopped and loaded the car with more samples, they headed back to base camp over a succession of hills, and Duke finally mentioned the obvious: "That fender was well needed, John. I'm being showered with dust."

"Yeah," Young said. "Didn't mean to do that. So am I." Later, he'd say that "it was just like we were watching rain," that dust "covered up the battery cover, and the instrument panel so bad that you couldn't read the 'Power Down' or 'Power Up' decals."

They pressed on to the lander, driving across subdued craters, up and over ridges. Along the way, they encountered a huge, oblong depression with a hole in its bottom. "I'm not going down that critter," Young declared. England asked Duke to shoot some movie footage of it. The astronaut said he couldn't—his hands were too exhausted and weak for the job. Instead, he suggested that Young circle the rover while he kept the camera steady on its post. With that, Duke invented the "LRV pan," a new and unanticipated use for the rover. They made two circles, then resumed their traverse back to the lander.

The longer they drove, the more they missed the fender. "Man, I am covered from head to foot with dust," Duke said. "Boy, those fenders really are useful, Tony. This one we lost in the back has resulted in us being—"

"Pretty dirty," Young said.

"—a double Pig-Pen."

FROM THE START, APOLLO 16'S THIRD AND FINAL EVA WAS MESSY. Houston did not recommend a fix for the broken fender, and the crew didn't attempt one. So, wherever it went, the rover kicked a plume of dust forward onto the men and the rover's electronics.

First stop: North Ray Crater. Unlike the day before, they moved quickly and easily across rolling plain that was relatively free of boulders. Their route took them up and around the rim of Palmetto Crater, named for Duke's home state of South Carolina. Small, as lunar craters go, it stunned nonetheless: the hole was nearly a half mile wide and more than four hundred feet deep.

From Palmetto's rim, they had a sweeping view of the rise to neighboring North Ray Crater, not quite two miles away. It looked like an easy ride. They dropped off the side of Palmetto, and soon Young was able to report the "rock population is almost nonexistent." Craters disappeared, too. The suddenly smooth, open space around them reminded Duke of West Texas. It didn't last: nearing the base of North Ray, they saw a battalion of boulders looming uphill. "Look at those rocks!" Duke cried. "Tony, there are some tremendous boulders on North Ray, and they get bigger as we go nearer them."

"I think we're starting to get into it right now, Charlie," Young said.

Mission Control, monitoring the rover's systems from Earth, told them that battery 2 was running hot, and a caution flag might pop up on the control console. If it did, they should just reset it and soldier on. "Now, before we get too far along," Young said, "let's study this thing and figure out a way to get up that rim without going through all the boulders in the world."

"I think up that slope looks to me to be the best," Duke said, indicating a route off to their right. He second-guessed himself. "Of course, it might be straight ahead might be best."

"Well, I don't see any rocks straight ahead," Young replied. "Let's try straight ahead."

The rover's nose ratcheted toward the vertical. They topped a series of false peaks, each time thinking they had arrived at the rim. Each time, Houston—knowing how many miles they'd need to drive to reach it—assured them they had not. When they finally neared the real thing, England asked them to look around in the crater for black rocks. Geologists on the ground still hoped for evidence of volcanic activity. If the meteorite that carved out North Ray hit hard enough, it should have tossed both white bedrock and black volcanic rocks around the rim. "Well, we can't see in the crater," Duke told him, but sitting up on top was "a biggie. The biggest one, Tony, is this ten- to fifteen-meter boulder that is on the rim, and it's blackish."

They stepped from the rover to peer into the crater. It was quite a bit larger than Palmetto: more than three thousand feet across and eight hundred feet deep. "Man, does this thing have steep walls," Young said. "Now, I tell you, I can't see to the bottom of it, and I'm just as close to the edge as I'm going to get." The pair spent more than forty-five minutes collecting rocks and raking regolith into sample bags before turning their attention to the black boulder. "It is a biggie, isn't it?" Young asked. "It may be further away than we think."

His instincts proved right. The rover's TV camera captured the astronauts dwindling in size as they ventured toward it. Eventually they all but disappeared behind a slight rise. "It was a pretty significant rock, so we started jogging that way," Duke told me. "The more we jogged, the bigger it got. When we finally got to it, we saw it was about sixty meters across and about thirty meters tall. We called it House Rock, because it was the size of a house, maybe even bigger." He added: "Depth perception is a problem on the moon."

It was dark, but not volcanic. Another breccia—a big one, to be sure, but not what mission planners expected. "We pounded on it with a hammer," Duke recalled, "and got a couple of chunks and took some nice photos."

Running late, Young and Duke hiked back, loaded the rover, and set off for the next station. Their route took them back down North Ray the way they came. "This is going to be something," Duke said at the top. As they had learned at Stone Mountain, driving downhill could be a white-knuckle experience. Young kept the brakes on all the way, and still they picked up speed. "Man, are we accelerating!" Duke hollered. "Super."

At the bottom, Young staked a claim. "We've just set a new world's speed record, Houston," he told England. "Seventeen kilometers an hour on the moon," or 10.54 miles per hour.

"Well, let's not set any more," the capcom scolded.

They pulled up at their next station so layered in dust that Duke couldn't read the meters to relay their settings to Mission Control. The ground crew reported that the LCRU was heating up, a signal that it, too, was coated. "Needs dusting," Duke confirmed. "Needs dusting, bad." England directed them to a boulder visible in the TV feed. "We'd like you to hammer on that rock a bit," he said. Studying the image, he added: "It looks like some soil on the south side, kind of underneath, might be permanently shadowed. You might take [a sample] and just put it in a bag." Shadowed meant that it had been protected from the moon's onslaught of cosmic rays and micrometeoroids, and would theoretically provide a time capsule of the regolith's state when the rock landed there. Duke walked to the boulder, dropped to his knees, and reached about a yard beneath its overhanging side to get a sample. He regained his footing and eyed his handiwork. "You do that in West Texas, and you get a rattlesnake," he told Mission Control. "Here you get permanently shadowed soil."

Their last station stop lay back near the lunar module. On the way, Young offered to ride the rim of Palmetto so they could get some pictures. They cruised at nearly nine miles per hour, approaching the crater on hard, smooth regolith and showered all the while by debris. When they hit a bump, a blast of it shot forward into their backs. As Wernher von Braun had rightly predicted in *Collier's*, it settled at once. And on the airless moon, none of it blew away.

They crested the rim, where Duke fired off a quick LRV pan, then started for the lander, less than a mile off, again under a rain of dust. "This is an absolutely great suspension system, Houston," Young radioed as they dodged small craters and cobbles. "You should see some of the things we've run through, and this baby just bounces right out and keeps right on going."

"We ought to see the old beauty when we top the rise here," Duke said, predicting another win for the navigation system. He paused as they ran over a rock the size of a basketball and summited the ridge. "There she is, John!"

"Well," England radioed, "that's good news."

"Don't run into our home," Duke advised.

Young chuckled. "Right, Charlie."

That last stop was brief. Later, after unloading their samples and tackling more overhead, Young drove about 265 feet from the lander and turned the rover around. He told Houston he was parked with the batteries pointed away from the sun, to keep their temperature down. "Okay, fine," England told him. The capcom asked for a final readout from the rover's meters. Young gave it to him. "That do it, Houston, for you?" he asked.

"Right," England replied.

Young unfastened his seat belt and climbed off. He and Duke had traveled 16.6 miles in LRV-2. That was slightly less than the Apollo 15 crew but still four times the total distance traveled by astronauts on the first three Moon landings. They had collected 209 pounds of samples. And they had gathered evidence refuting a widely held theory about the genesis of that part of the moon's surface. All impossible without their trusty steed.

With a lot to do and less than five hours before their scheduled liftoff, the pair hurried to shuttle their sample bags up the ladder. Young took a last photo "of the old rover sitting there," adding: "Boy, that's a good machine."

"Yeah, it's an incredibly good machine," Duke said.

Young turned to his partner. "Now, we got some work to do here, boy. You're all dirty."

"You ought to see your back," Duke replied. "I couldn't have gotten any dirtier than you."

The task of dusting each other off took several minutes. "Well, the message is clear," Young said as Duke attempted to beat the dust off his arms.

"What?"

"Don't lose the fender off the rover."

Minutes later, they were in the lander, the hatch locked behind them. Then, just under three days after landing, *Orion* blasted free of its descent stage. Ed Fendell, guiding the rover's TV camera, was able to keep up with it for six seconds as it soared into space.

"There we go!" Duke cried. "What a ride!"

60

TWO MISSIONS, TWO RUNAWAY SUCCESSES. THE FIRST TWO rovers had covered nearly thirty-four miles of the moon's surface, had clambered over and around every obstacle in their paths, and had betrayed only minor gremlins. The steering anomalies had sorted themselves out. John Young's loss of rear-wheel drive appeared to be a clear case of operator error. In a postflight technical debriefing, he and Charlie Duke had nothing but praise for their machine.

Duke: "Very comfortable, riding in the rover."

Young: "Beautiful suspension system. If it hadn't been, we probably would have walked a long ways."

Duke: "Really outstanding. You could have three wheels off the ground at once, and the thing would just recover smartly, and it was just beautiful."

Young: "On at least three different occasions, we bounced up in the air and came down on a rock, which we were passing over, and it didn't seem to affect the operation in the slightest."

The LRV project team at the Marshall Center saw little reason to fix what wasn't broken as it prepared the third rover for Apollo 17. For a while, in the spring of 1971, it had entertained a GM proposal to add a remote function to the third rover, enabling Mission Control to keep it in service after the astronauts finished their work. Inspired by Lunokhod-1, Sam Romano pushed the idea up and down NASA's chain of command—which, considering the headaches and anxieties of the previous two years, might be the very definition of chutzpah. In the end, though, NASA decided against the idea, so the machine folded and loaded into the lunar module *Challenger* was close to a clone of its predecessors. Its most obvious modifications: stops to discourage its fender extensions from breaking off or sliding too far. The Marshall labs, which designed and built the stops, hoped they'd prevent another dust storm.

Maybe it was inevitable, the way that turned out. Perhaps the universe's sometimes perverse sense of humor dictated that before LRV-3 had started its first run—on what was supposed to be, by far, the lunar rover's most demanding test, and the Apollo program's capstone expedition to the moon—mission commander Eugene Cernan would walk too close to the right rear wheel with a hammer sticking out of his shin pocket. And that a moment later he'd groan, "Oh, you won't believe it."

It was a near-exact duplication of John Young's misstep, only worse: Cernan had been out of the *Challenger* for only an hour and forty-one minutes; he and his lunar module pilot, Harrison "Jack" Schmitt, had barely started their mission. And a look around their

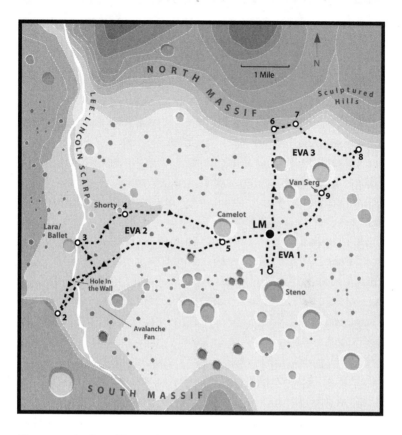

The mountain-ringed Taurus-Littrow Valley, where Apollo 17 was to close out America's rover project and the Apollo program in fitting style. (KEVIN SWIFT)

landing site underscored the essential role the rover was to play in it. They had set down near the center of the Taurus-Littrow Valley, on the eastern fringe of the Sea of Serenity, otherwise known as the man in the moon's left eye. A couple of miles away, the South Massif formed one of the valley's walls, rising to a peak 7,500 feet above its base. They were due to go there. At the mountain's foot, the spilled debris of an ancient avalanche fanned halfway across the valley floor. They were to sample it. A roughly equal distance to the north stood the 6,600-foot North Massif, and to its east, a jumble of lower peaks called the Sculptured Hills. Both were planned destinations.

The Taurus-Littrow promised adventure. Floored in volcanic basalts overlaid by a thick blanket of regolith, it contained three different moonscapes within rover range of the lunar module: mare, ancient highlands, and what were believed to be younger volcanic fields. Of further geologic interest, the valley's opening into the Sea of Serenity, off to the west and about four and a half miles wide, was partially blocked by the Lee-Lincoln Scarp, a fault line that crossed from South Massif to North, and in places stood 250 feet tall. They were scheduled to explore its top.

Cernan recognized that dust was no minor inconvenience. He'd seen what had happened to Young and Duke on a relatively short drive and knew that on the marathon trips planned for this mission, he and Schmitt would squander precious time cleaning the rover and themselves. They wouldn't be able to drive as fast, either, further cutting into what they could accomplish. After a moment of taking all this in, Cernan broke the bad news: "Lost a fender."

"Oh, shoot!" Schmitt understood what it meant, too.

"I'm going to have to try to get that fender back on," Cernan said, studying the damage. Sliding the extension back into position wouldn't work—he had wrecked the flange that held it in place. He turned instead to duct tape. Trouble was, there was already so much dust on his gloves and the fender that he couldn't get it to stick. It took several tries before he managed an unsteady repair.

"Bob, I am done," he told their capcom, physicist and astronaut Robert Parker. "It's not too neat, but tape and lunar dust just don't hang in there together." He jiggled the fender, testing its strength. It wasn't great. "Keep your fingers crossed," he said.

Spoiler alert: for all of duct tape's virtues, it has limitations.

Murphy's law is named for an aerospace engineer, you know.

The first drive was a short one, which was a lucky break. They would head less than a mile south to the rim of Steno-Apollo Crater, where they would sample pieces of the valley's basalt floor exhumed from beneath the regolith. They would stay thirty minutes, then drive back. Along the way, they'd plant explosive charges for a seismic profiling experiment.

The two astronauts had studied the valley's topography, practiced flying and driving it in simulators, and rehearsed their excursions. Yet once under way, the pair found—like those who'd come before—that they recognized little of the ground they crossed. Parker tried to talk them to Steno-Apollo: "It should be up on the top of a little bit of a rise that you see coming up there," he said. And: "You ought to be in the vicinity of some very large boulders."

"Houston, there are certainly a lot of big boulders," Schmitt said.

"That doesn't look what I thought Steno looked like," Cernan said of a crater they were nearing.

Parker: "Steno ought to be right at your nine o'clock there, Gene."

Schmitt: "Well, it doesn't look real familiar, Bob, as far as Steno's concerned."

On the west side of Steno, Parker said, they should see two clusters of boulders and try to sample between them. They saw no such boulders. "It's hard to follow that that's where we are," Schmitt said. "I'm not sure."

"No," Cernan said. "Me, neither." He thought he saw a crater he recognized—Emory—on a nearby rise. It wasn't Emory, but he headed toward it. After a while, they came to a field of small boulders, and Cernan made a command decision. "We better park in this boulder field here," he said.

They climbed off the rover six hundred feet short of Steno-Apollo's rim. Schmitt paused to plant an explosive charge. They would place eight of them during their three days in the valley. The shocks created by their detonation, after the astronauts left, would be read by a seismic receiver, and supply all manner of arcane data about the lunar crust. "Get your hammer," Schmitt told Cernan. "We're going to need it."

On the earlier Apollo missions, Mission Control and its advisors from the Geological Survey had spelled out the crews' scientific duties in great detail. This time NASA had an expert on the ground. Schmitt was the first and only bona fide scientist on an Apollo flight, with a doctorate in geology from Harvard University. He had served as a project chief with the Geological Survey in Flagstaff before he was selected as an astronaut, and had been instrumental in developing his fellow spacemen into competent field geologists. He'd originally been scheduled to fly on Apollo 18. After its cancellation, pressure from the scientific community spurred NASA to move Schmitt, a thirty-seven-year-old bachelor from New Mexico, to this last lunar mission.

He and Cernan walked among the boulders scattered around the top of a small crater, sixty-five feet wide and a dozen feet deep, until Schmitt chose one for sampling. Cernan beat its edges with the hammer while his partner gathered the chips in a bag. After a great deal of pounding, Cernan managed to break loose a slab the size of his hand. They circled the perimeter, sampling another interesting boulder, shooting panoramas, collecting rocks from the surface, raking the soil. Before they left, Cernan checked his repair. The duct tape was holding.

Three-eighths of a mile from camp, they stopped for Schmitt to place another explosive charge. They circled it for an LRV pan, then set off again in a dash—and Schmitt noticed dust pelting him from the rear. "I think you have lost a fender," he said. "Look at our rooster tail." Cernan couldn't see over his right shoulder to the fender, but in the machine's shadow to his left, he could make out the right rear wheel hurling a thick tongue of dust over their heads. Some of it overshot the rover to land on the surface in front of them. "Oh, boy. That's going to be terrible," he said. "Look at the LCRU." Dust already lay thick on the box of electronics linking them to Houston. He could barely read the console.

Back near the lunar module, Cernan used the rover to lay out a grid in the regolith for a new experiment measuring the surface's electrical properties. The navigation system enabled him to create perpendicular lines that guided the crew's positioning of the experiment's long wire antennas. It put the rover to another creative use, but also buried it in still more dust. Cernan worried. If they attempted the next day's long drive without a fix, he told Houston, they'd be in trouble. "I can stand a lot of things," he said. "But I sure don't like that."

He drove back to the lander and started the long process of clearing dust from the rover's electronics. "Give me a yell when you need a spell there," Schmitt told him.

"What? Dusting?"

"Yeah."

"Well, I need a fender," Cernan replied. "That's what I need. Figure out something we can make a fender with."

"How about one of the others that's not as critical?" From one of the front wheels, Schmitt meant. Cernan wouldn't dare. They weren't supposed to come off unless they broke, and one busted fender was plenty, thank you.

It was when they dusted themselves off before reentering the lander that the full impact of the fender's loss came home to them. Their suits were filthy. Dust filled their pockets, sweatered their backpacks, and had insinuated itself into the complex aluminum rings locking their gloves and helmets into place. Those rings could take only so much dust before they started seizing. And the stuff was both abrasive and so fine that it smeared: wiping it from their helmet visors scratched the delicate coatings, which threatened to play hell with visibility. Worst of all, it was only a matter of time before the rover's dust-covered electronics melted down.

The fender's loss threatened to compromise many of NASA's aspirations for the last, grandest Apollo mission to the moon.

61

IT HAD OPENED WITH A FLOURISH, WITH THE PROGRAM'S ONLY night launch, in the wee hours of December 7, 1972. The fiery tail from Sonny Morea's F1 engines, half again as long as the rocket itself, could be seen hundreds of miles away as the last manned Saturn V shot skyward. Four days later, *Challenger* separated from the command module and dropped to the surface, leaving *America* and its pilot, Ronald Evans, in lonesome orbit.

Cernan, thirty-eight, was a capable, seasoned choice as commander. He'd grown up in the Chicago suburbs and entered the navy through Purdue University's ROTC program. He made two hundred carrier deck landings before he was selected as an astronaut in the third group, with Dave Scott, and had been in space twice before: on Gemini 9A, during which he performed a spacewalk of more than two hours, and with John Young on Apollo 10, the run-through for the first landing. Cernan and his commander on that mission, Tom Stafford, had descended in their lunar module to within 47,000 feet of the moon's surface. He had worked hard for the chance to return and to finish the descent.

That night, the commander lay worried in his hammock in the lunar module, while a small army studied the fender issue in Houston. The mission's Flight Crew Support Team went over everything stored in the lander, found on the list a few useful ingredients, and used them to cobble together a possible repair. John Young donned a pressure suit to walk through its installation with one of the rover test units. It seemed to work.

Shortly after Mission Control woke the crew with Richard Wagner's "Ride of the Valkyries," Young came on the radio. "Hey, we spent some time on this fender problem and worked out a pretty simple-minded procedure," he said. They were to take four laminated pages from their Geological Survey map package, and duct tape them into a single sheet measuring fifteen by ten and a half inches. Then they were to scavenge the clamps from the lunar module's pair of hang-anywhere task lights, and use those to pin the maps to the fender. The fix recalled a similarly makeshift repair during the beleaguered Apollo 13 mission, in which Houston engineered a carbon dioxide scrubber from duct tape, tube socks, the paperboard cover of a flight

All through the 1960s, GM's lunar vehicle designs relied on flexible frames and six or more wheels. Here, in a painting by company designer Norman James, the mobile laboratory it brainstormed with Boeing is poised to conquer the lunar wastes, along with an open "moon jeep." *Norman J. James and General Motors*

The leaders of GM's lunar mobility team—(FROM LEFT) Sam Romano, Greg Bekker, and (in the driver's seat) Ferenc "Frank" Pavlics—pose with their mobility test article in 1966. The oversized machine remained earthbound, but pieces of it, including its wire-mesh wheels, would find their way to the moon. *Romano family*

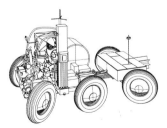

(ABOVE RIGHT) GM's local scientific survey module, or LSSM—a shrunken version of the mobility test article, and a step closer to the rover's final design. *Norman J. James and General Motors*

Boeing/GM's chief competition came from Bendix, whose concepts were short on elegance but simple, sturdy, and well-tested. Here, technicians pilot a demonstrator that had much in common with the company's rover proposal. *Bernard Pilon*

Grumman's remarkable rover prototype prepares to dive into a crater on the company's Long Island moonscape. The craft was limber and surefooted, but perhaps just a touch too radical for NASA. *Bruce H. Frisch/Science Photo Library*

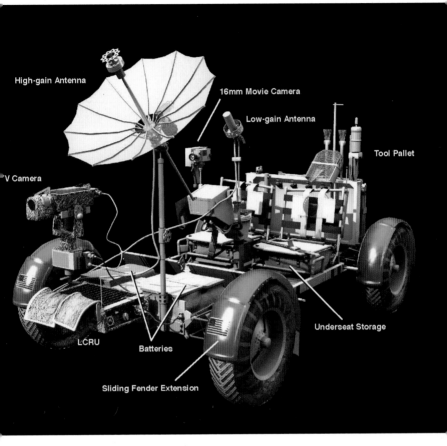

High-gain Antenna

16mm Movie Camera

Low-gain Antenna

Tool Pallet

V Camera

Underseat Storage

LCRU

Batteries

Sliding Fender Extension

The winning concept from Boeing/GM was the most conventional, in automotive terms, of the four considered for Apollo, with its powerplant up front, its crew seated side by side amidships, wire wheels that emulated pneumatic tires, and a "trunk" at the rear for tools and rock samples.
Don McMillan

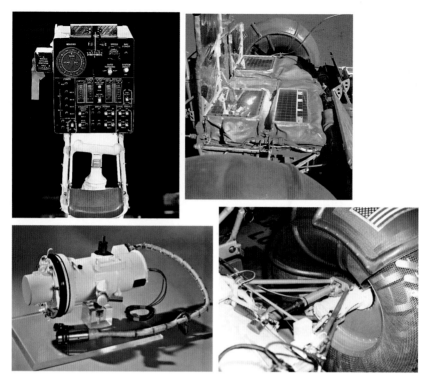

Elements of the Boeing/GM design: the vehicle's T-bar hand controller and an instrument panel that showed the way and monitored the rover's systems; silver-zinc batteries topped with mirrored radiators (here shown with their protective covers raised) to regulate their temperature; the compact, light, and hermetically sealed assembly that housed the motor and transmission in each wheel hub; and a simple, strong suspension and exotic wheels that met the demands of the broken lunar surface. *Clockwise from top left: NASA, NASA, Pavlics family, NASA.*

GM's struggles with the rover sparked NASA's worry and presence. Here Frank Pavlics, right, explains a detail of the 1G trainer to Saverio "Sonny" Morea, left, and astronau Jerry Carr while they visi the automaker's Santa Barbara lab. *Pavlics fami*

pollo 15 commander Dave Scott, right, and lunar module pilot Jim Irwin drive the rover alongside New Mexico's Rio Grande Gorge, March 1971. The U.S. Geological urvey cobbled together the sorely needed training vehicle for less than $2,000. *NASA*

The Geological Survey's Laszlo Kestay stands beside one of dozens of craters at Cinder Lake near Flagstaff, Arizona, a half century after lunar analogs were created there for astronaut training. *Earl Swift*

Encased in their extravehicular mobility units, Dave Scott, right, and Jim Irwin take the 1G trainer for a spin at the Kennedy Space Center, May 1971. Emulating much of the actual rover's behavior, the trainer enabled crews to fine-tune procedures for exploring the moon. *NASA*

Scott, second from left, and Irwin, far right, check out LRV-1 with Kennedy engineers and fellow astronaut Bob Parker, far left. The Apollo 15 crewmen are wearing space suit gloves, to get a feel for the machine as they'll actually experience it. *NASA*

Dave Scott, right, checks the folded rover's fit in the lunar module *Falcon* during a test at Kennedy. *NASA*

The Apollo 15 rover in its stowed position aboard *Falcon*, May 1971. Nose down and belly out, the machine is shielded by the underside of its floorboard. *NASA*

pollo 16 commander
hn Young practices
eploying the rover,
ing one of the
ie-sixth gravity test
odels, January 1972.
y the time he and
ther moonwalkers
ached the moon,
iey'd so rehearsed the
rocess it was almost
ivoluntary. *NASA*

Apollo 15's Saturn V rides out a lightning display on the launch pad at Kennedy Center on the night of July 25, 1971. The next morning, the rocket blasted off for the moon with LRV-1 tucked into the lunar module *Falcon*, which was housed in the conical shroud below the capsule.   *NASA*

im Irwin tidies up following the rover's inaugural voyage, Mount Hadley towering lmost three miles high in the background. Note the sample collection bags hanging rom the rover's rear and the antennas—umbrella-shaped high-gain on the left, pacifier- haped low-gain on the right—pointed at Earth, directly overhead. *NASA*

NASA brass declared that LRV-1 "made the mission" for Apollo 15—and in more than seventeen miles of exploration, it did just that. Here, the rover is parked near *Falcon* on the Hadley Plain, ringed by landforms without earthly equivalent. *NASA*

John Young lets 'er rip during Apollo 16's "grand prix" in the Descartes Highlands, in a frame grab from the 16-millimeter film shot by Charlie Duke. The short clip offered engineers back home a glimpse of the rover's handling under way. *NASA*

…ag in hand, John Young collects samples during the Apollo 16 crew's third EVA.
…he rover's tailgate—its tool pallet—hangs open in this view. Black stripes on its
…heel hubs helped engineers track movements while reviewing the grand prix
…otage.   *NASA*

John Young uses a sighting device to aim the rover's high-gain antenna, which carried TV signals home from the moon. This image, shot during the crew's third EVA, shows that the right rear fender has lost its sliding extension. *NASA*

Charlie Duke circles the front of the rover, late in Apollo 16's mission. His grimy suit testifies to the dust heaved on the crew after the fender mishap. *NASA*

Apollo 17 commander Gene Cernan takes LRV-3 for a maiden cruise shortly after its release from the lunar module on December 11, 1972. The rover appears even more minimalist than usual, as it has not yet been fitted with its antennas, TV cameras, LCRU, and tools. *NASA*

Apollo 17's Harrison "Jack" Schmitt leaps into his seat during an EVA in the Taurus-Littrow Valley. The astronauts' stiff space suits required them to use their seat belts to cinch themselves into a seated position. *NASA*

Apollo 17's well-traveled LRV-3 parked at Station 2 near Nansen Crater on December 12, 1972. The stop, 4.72 miles from base camp, was the farthest point any Apollo crew ranged from its lunar module—and arguably, the outermost limit of man's explorations on the moon or anywhere else. *NASA*

ck Schmitt runs, sample scoop at the ready, at Station 5 near Camelot Crater, late
Apollo 17's second EVA. This is a color photo, which attests to how leached and
onochromatic the moonscape appeared   *NASA*

ene Cernan checks out the rover at base before setting out for the North Massif. The
ew's makeshift fender repair, curled and battered, is showing the abuse it absorbed in
ore than twelve miles of driving the previous day.   *NASA*

Jack Schmitt works the boulders at Station 6 on the North Massif during Apollo 17's third and last driving exploration. The rover's tilt offers a hint of the slopes at the site, which are otherwise hard to read. *NASA*

Greg Bekker had some fun in a watercolor he painted six years after retiring: it shows John Young in the Apollo 16 grand prix, zooming past the remains of a rover conceived by Polish science-fiction novelist Jerzy Zulawski in 1901. *M. G. Bekker family*

plan book, and plastic bags scavenged from inside the damaged space-craft. "The beauty of it is you're only spending about two minutes in the clamping operation," Young said. "What do you think?"

"I'd sure like to give it a stab," Cernan answered. "The only hooker is I hope that tape holds the fenders together well enough."

"Roger," Young replied. "One of the things, when you're taping the pages together, that you want to be careful of is that you make sure and get the air bubbles out, so when you get in the vacuum, it doesn't open up by itself." He would watch them on TV, he promised, and could offer advice as needed.

When the astronauts stepped onto the surface to begin their second day on the moon, they carried the taped-together maps and clamps. They loaded up the rover before Cernan moved to the right rear wheel, replacement in hand. "It's just a flapper," he said of the assemblage. "Sure isn't stiff like I want it to be." When Schmitt pressed the maps over the fender's curving front portion, however, they became a surprisingly sturdy half tube. Cernan lifted its rear edge to check its clearance. "Well, it'll stop some of it," he said. "If it stays on." They adjusted the fit, straightening the paper and shifting the clamps. The new extension wasn't pretty—it jutted straight back and looked every bit the act of desperation it was—but it was better than nothing.

"Okay," Schmitt said. "Let's go."

Nearly nine years before, Cernan had been one of four rookie astronauts to dine with Wernher von Braun at a Houston restaurant. The rocketeer had predicted that Apollo's moon men would have much to do on the lunar surface. "Lindbergh didn't fly the Atlantic to get to Paris," he told Cernan. "Gene, you are going to need mobility. We will provide a car."

"Whoa," Cernan thought. He didn't voice how nutty that sounded.

Now look: their second excursion took them southwest across the valley's dark and pitted floor toward the avalanche splayed at the

Gene Cernan's replacement fender for LRV-3, designed overnight by a team in Houston. Fashioned from Geological Survey maps, duct tape, and clamps, it wasn't much to look at—but it worked. (NASA)

South Massif's base. They paused so that Schmitt could drop another explosive charge without leaving his seat, slowed for pictures on the blocky rim of two-thousand-foot Camelot Crater, and braked hard for small craters waiting in ambush. Otherwise Cernan kept the stick at its limit.

Before them, the Lee-Lincoln Scarp crossed the valley high and steep. Near its collision with the South Massif, the scarp dipped, and that notch coincided with the top of the avalanche fan. The first geologic stop of the day was through that gap, which the crew had named Hole in the Wall. They stopped briefly to get an "LRV sample." Dave Scott's subterfuge to collect the seat belt basalt had inspired the creation of a long-handled sampling tool that Schmitt could use without unbuckling his own belt. Three miles into the drive, a second LRV sample brought more soil and rocks aboard.

From a distance, the avalanche appeared much lighter than the valley floor. Now, as they reached its edge, the contrast faded. "Oh, man," Cernan said as they searched for a smooth route up. "What a trip this is going to be." Then he found, as had the two drivers before him, that "this rover doesn't know it's going up a hill." The going got very steep in spots, he reported, but their machine, throwing all of its single horsepower into the climb, appeared to be unbothered. At the crest, Cernan steered them into Hole in the Wall. "Let me tell you," he radioed, "this is quite a rover ride."

On the scarp's far side, they descended into a wide, shallow trough between the South Massif's base and the surrounding moonscape. Their target, Nansen Crater, was not the result of an impact but merely a low spot in the trough. Trails left by rolling boulders scored the mountainside above it; now those rocks rested near Nansen's edge. Sampling them was a priority because they clearly came from outcrops high overhead. They offered pieces of mountaintop without a long and treacherous climb.

Cernan and Schmitt pulled up beside a boulder fragmented into a handful of large blocks. They stepped from their machine having established a new benchmark of lunar exploration. The rover's console indicated they were 7.6 kilometers, or 4.72 straight-line miles, from base camp—very near the practical limit of their range. NASA's walk-back constraints permitted them to drive farther from the LM; they had enough air and water on their backs to do so. But if they ventured past this point, they wouldn't have time to do much geology.

This far out, they couldn't dawdle when Houston told them it was time go, either. The drive to Nansen had taken sixty-one minutes. If one of their backpacks broke down, they would have to buddy up on the cooling water in the surviving pack and tap into an eighty-minute emergency oxygen supply in the crippled one. The arrangement gave them just enough time to drive back to camp, lock themselves into the lander, and pump it full of air.

Earth, captured in a photo Gene Cernan snapped over a boulder fragment at Station 2—and a view that brought home just how far he and Jack Schmitt had ventured. (NASA)

While planning the rover missions, NASA did some hand-wringing about how to handle failure of the rover *and* a backpack. It arrived at no solution: in that worst-case scenario, an astronaut would likely die. The agency's insistence on redundancies all but ensured Cernan and Schmitt against such a fate. Even so, they were out on the edge of the program's margins. The scarp blocked their view of most of the Taurus-Littrow Valley, and down in the trough, they couldn't see far in any direction—the depression's rim tightened the horizon to just a few yards. "Man," Cernan said, "we've got to be a million miles away from the LM."

No surprise, that it felt that way. In its Apollo 17 mission report, NASA would note, with characteristic understatement, that Cernan and Schmitt had reached "the greatest radial distance any crew has traveled away from the lunar module on the lunar surface." Put in slightly less guarded terms, they were at the farthest point to which man has ventured in the flesh, on any world. Here they'd leave humankind's outermost footprints.

They split up to examine and sample different boulders and called each other over now and then to show off a find. They worked together to break chips from the rocks. Houston offered them an extra ten minutes, to be subtracted from their third stop of the day. They took it. A few minutes later, they gathered pieces from one of the smaller boulders in the group. It contained a white clast, a knot of bright crystal in the midst of darker breccia. "Man, that's a prize," Cernan said. He hammered free several fragments. One was found to be, at an estimated 4.6 billion years old, the oldest rock ever brought back from the moon.

Then, all too soon, Houston was moving them on.

Back through the Hole in the Wall, they had a panoramic view of the valley below, though they couldn't quite make out the lunar module. Cernan steered the rover away from the mountain; they rolled downhill on the avalanche debris, then veered toward the scarp. It was less a wall than a chain of overlapping billows, some of their sides steep but nowhere near vertical. Station 3 lay alongside Lara Crater, which was drilled into one of them.

Cernan casually asked whether Schmitt had noticed how fast they had been going: "What was it, seventeen and a half or eighteen clicks we hit coming down?" He was speaking in kilometers—translated, 10.9 to 11.2 miles per hour. Schmitt just laughed, and Cernan allowed later that he might have been exaggerating, but the remark, preserved in transcripts of their radio chatter, earned Cernan and LRV-3 the moon's land speed record.

Alongside Lara Crater, they dug a trench and sampled its floor and sides, took cores, measured the site's gravity. They gathered rocks. They took photos. Cernan packed soil into a vacuum-sealed can that NASA planned to leave unopened as a sort of time capsule, pending future advances in analytical technology. The agency finally uncorked it in 2019.

Schmitt lost his balance and pancaked several times onto the lunar surface, prompting their capcom to remark that "the switchboard here . . . has been lit up by calls from the Houston Ballet Foundation, requesting your services for next season." At that, Schmitt attempted a balletic leap, and again went sprawling. More than forty-seven years later, in February 2020, the International Astronomical Union, a Paris-based congress of space scientists from around the world, approved renaming the Station 3 site Ballet Crater.

Again edging close to their walk-back limit, the astronauts took off for Station 4. They dropped off the scarp and across the avalanche debris. The schedule called for them to make an intermediate stop for a quick sample. Houston canceled it, but they stopped anyway. Accustomed by now to moving in their suits and handling their sampling gear, they knocked it out in thirty seconds.

Out near the toe of the avalanche fan waited Shorty Crater, 360 feet wide and 45 deep. Their assignment: samples from the crater rim and panoramic photos. So, while Cernan dusted the rover's electronics, Schmitt prepared to take pictures. As he did, the ground at his feet caught his eye. After two days in the moon's monochromatic, mousy gray-brown, its color jolted him: pale orange. "Oh, hey," he said. "Wait a minute."

Cernan paused in his dusting. "What?"

Schmitt checked whether the orange foil cladding the TV camera was reflecting onto the ground. It was not. "There is orange soil!" he cried. In Houston, Ed Fendell swung the TV camera toward him. The color was visible on the monitors at Mission Control. "It's all over!" Schmitt said. "Orange!"

"Don't move it until I see it," Cernan said, hurrying over. "Wait a minute. Let me put my visor up." The sun visors that slid down over their helmets' glass were reflective gold. Sliding his out of the way changed nothing. "It's still orange!"

Schmitt didn't wait for instructions. "I've got to dig a trench, Houston." Shorty Crater had been the focus of geologic speculation before the mission. Its dark appearance in photos shot from lunar orbit suggested a volcanic vent rather than an impact crater. The presence of orange soil certainly bolstered that impression. It would turn out that the orange soil *was* volcanic, after a fashion. Tiny spheres of volcanic glass, the product of a fountain-like eruption billions of years back, colored the soil. Covered in turn by the magma that formed the valley's floor, then by the avalanche off the South Massif, the glass had been blasted back to the surface by the meteorite that created Shorty—which was your standard impact crater after all.

But the astronauts didn't know that. Schmitt, working on his trench, saw all the signs of a vent. "I think we hit one of those things we've got to reconsider on, Houston," he said. Capcom Bob Parker had bad news: they had to go. They hadn't closed the distance to base camp enough to buy much time at Station 4. Cernan and Schmitt hurried to collect soil from the trench and took a deep core sample. The core tube came out of the ground colored the same orange. Houston was pleased. "Quite a station, men," Parker told them. "[And] we thought Station 2 was a good station!"

They rolled off the avalanche fan and back onto the valley's undulating floor, then struck east for home. Just shy of a mile along, they stopped to plant another charge and took an LRV pan. Blinded driving into the sun, Cernan zigzagged to the north and south, tacking. Their final stop, back at Camelot Crater, lay less than a mile out. Houston gave them twenty-five minutes, and they hurried through their sample collection before Parker instructed them to "leave immediately, if not sooner."

As Cernan unloaded the rover back at base camp, he examined the replacement fender. It was scoured and stained by abrasive regolith but remained intact and firmly attached. "Hey, congratulate Jose"—John Young—"on that fender, will you?" he radioed Houston. "I think he just saved us an awful lot of problems; he and whoever else worked on it."

In Neptune City, New Jersey, the repair moved the president of the Auto Body Association of America to declare Cernan and Schmitt "damn good body and fender men" and to award them honorary life-time memberships in his organization. "We're delighted to see that when something like this happens on the moon, they had the ingenuity to put it all back together," Reg Predham said. "I think General Motors and other carmakers ought to take note—250,000 miles away on the moon they can make repairs, but we can't get parts in New Jersey."

Later, in the lunar module, Cernan and Schmitt went over the day's accomplishments with Joe Allen, who'd taken over as capcom after the EVA. They had covered 12.49 miles on the excursion, by far the longest single drive in the rover's brief history. They had spent seven hours, thirty-seven minutes outside the lander, another record. And they had pushed the edge of a crew's range of exploration, at almost five miles from camp. Surprisingly, their metabolic rates had been lower than on the day before. "We spent a lot of time driving today, and a lot of time working yesterday," Cernan pointed out.

"That's true," Allen said. "We can ask for the metabolic rate of the rover. I bet that is pretty impressive for today."

"Well, don't get me wrong," Cernan said. "Driving that rover is soft, but I'll tell you, it keeps your attention."

Schmitt: "It keeps the passenger's attention, too!"

"That's a super machine to drive, though, Joe," Cernan said. "If you had enough time, you could really learn to take it all the way."

"Geno, was it spraying dirt at you today?" Allen asked. "Could you notice that you still missed the real fender . . . ?"

"No, sir. I don't think we missed it at all."

"Fact is," Schmitt said, "we're recommending a design change."

"That'll be for next year's models," the capcom replied.

62

THEIR LAST DAY'S EXPLORATION OF THE MOON WOULDN'T BREAK any records for time or distance, but that said, Wednesday, December 13, 1972, promised to be full. Cernan and Schmitt would strike north this time to explore the lower reaches of the North Massif, then turn east to the Sculptured Hills. The "nominal" plan called for five scientific stops. It was early afternoon in Houston when Mission Control roused the astronauts from sleep. The ground team would work deep into the night.

The rover's battery 2, which ran warmer than expected on the first EVA, had run so hot on the second—125 degrees Fahrenheit—that it tripped a flip-up warning flag on the console. The Marshall Center assured Houston it could reach 140 degrees without a problem, and the drive back from the scarp had ended with the temperature a few degrees below that threshold. Now, as Cernan readied the rover for the trip north, he found that battery 2's temperature didn't register. The gauge had quit. He and Schmitt were still loading the machine when the warning flag again popped up. "Already, huh?" Cernan mused. With Mission Control's blessing, they carried on.

Twelve hundred yards from camp, they stopped for a rover sample, Schmitt directing Cernan to a spot he found encouraging. Farther north, the going turned rough, with blocks and craters abundant. Cernan, who drove flat out to preserve time, slammed the rover into both. Up ahead, the North Massif rose pale gray and mammoth against the black sky. On its flank, their target came into view: a boulder, or closely bunched pieces, a good way above the valley floor—Station 6. "Can we get up there?" Schmitt asked.

"It's awful high," Cernan replied. "We'll see."

Near the foot of the mountain, Cernan saw a way up, on an oblique track across the massif's face. As they climbed, the Station 6 boulder drawing near, they could see the chain of craters left by the bouncing rock as it had crashed down the massif. "We'll be looking right up that track," Cernan said. He asked his partner how he was doing. Fine, Schmitt answered, "until I looked down here and saw the slope we're on."

They stopped just west of the boulder, on its shady side, which wouldn't make for the best TV picture. The spot was far from level, too, but Cernan couldn't find a better place to park. Schmitt eased off the rover carefully, lest he tumble down the mountain. "You want me to block the wheels?" he asked Cernan, who was struggling to pull himself out of his seat. He added: "You got the brake on, I hope."

The incline complicated the simplest of tasks. Even dusting the rover was a challenge: When Cernan leaned over the machine from its uphill side, he toppled onto it. When he tried from below, he couldn't reach the batteries. Their heavy backpacks threw off their centers of gravity, making even picture taking a shaky balancing act.

Parker radioed that Charlie Duke, sitting at Mission Control, was comparing the Station 6 boulder to House Rock. In truth, it was a fraction the size, and broken into five pieces. "It's very crystalline," Schmitt replied. "I'll tell you, it's not a breccia, not like House Rock. Not to take anything away from House Rock." At first, he took it for igneous—some sort of anorthosite, perhaps, with bits of other rock fused into the mass. As Schmitt studied the boulder's biggest piece, however, his thinking shifted. It looked like two forms of breccia, after all, one of which had been molten when it fused with the other.

Cernan, over by the rover, noticed a dent in one of the tires. It was "a little golf-ball-size, or smaller, indentation in the mesh," he said, adding: "Doesn't hurt anything."

"That sounds like a dented tire," Parker agreed. It was the only damage sustained by Pavlics's wire-mesh wheel on the moon, despite some daredevil driving and countless collisions with rocks and craters. And while under way, the dent was undetectable; as Cernan said later, "you wouldn't know it was there."

Station 6 now yielded its most intriguing find. Schmitt moved away from the boulder to sample soil from one of the craters in the rock's wake, and pulled up a 5.6-ounce rock known as Troctolite 76535. NASA would regard it "without a doubt" as "the most interesting sample returned from the moon"—a colorful, well-preserved jumble of crystals, apparently assembled while part of slow-cooling magma, deep underground and early in the moon's history. Geologists studying its properties would find in them evidence that the moon once had an active magnetic field.

As they prepared to leave, Schmitt couldn't safely climb into the rover—it was tilted so far to starboard he might bounce out and keep bouncing all the way down the North Massif. Instead, he walked downhill to meet Cernan on more level ground. Mission Control had planned Station 7 for a short distance to the east but let the crew pick its exact location. They didn't stay long. Station 8 took them farther east to the sides of the lower Sculptured Hills. Again Parker told them to select any spot that looked promising. Cernan ventured only to the edge of the range's front before he stopped, this time with the rover's nose uphill, so they could both get on and off. Circling the rear end, he discovered that the taped-together maps had slipped free of the inboard clamp. The laminated paper had taken a beating. "Our fender's beginning to fade," he told Schmitt.

His partner started uphill to examine a small boulder that, unlike those around it, was not buried in regolith but sitting on the surface, as if it had rolled there. "If you need me," Cernan told him, "I'll come up there, because I think that may be worthwhile."

Schmitt had a better idea. "I'll roll it down to you," he said. "Are you ready?"

"I'm not sure I am," Cernan replied, "but go ahead." Schmitt pried up the rock with the toe of his boot and kicked it downhill. It thumped toward the rover, then swung into a hollow and stopped. "*I* would roll on this slope," Schmitt told the boulder. "Why won't *you*?"

"Five-sixths gravity that's missing," Cernan suggested. They hammered chips from the rock, discovering a trove of crystals under its coating of dust. They raked additional samples. While Schmitt dug a trench for still more, Cernan fixed the fender. Then they were again on the move.

Within minutes, Schmitt noticed dust hitting him from the rear. "Starting to rain again," he told Cernan. "Starting to sling dust. I wonder if we've lost our fender." Cernan didn't think so—he had made sure the clamps were screwed tight. But he guessed the taped maps, sandblasted with lunar dust for two days, might be breaking down.

The route to Station 9 took them through a maze of boulders, forcing Cernan to drop speed and steer a wandering course. When they stopped at the small, worn Van Serg Crater, a glance confirmed that their taped-together repair neared its end. The fender's rear edge had so curled and warped that it no longer blocked the spray coming off the wheel. Still, it had saved the mission. As Laszlo Kestay told me while we toured Cinder Lake, "That was one of the best uses ever of a U.S. Geological Survey map."

On photos taken from orbit, Van Serg looked a lot like Shorty, and the astronauts wondered whether they'd discover more orange soil near its rim. They found none. In their samples and cores, they found no basalt, either. The meteorite that created Van Serg apparently hit a thick patch of regolith and burrowed short of the valley's subfloor. The rocks around the crater's rim were dust fused into "instant rock" breccias.

Exhausted, Cernan and Schmitt posted no objections when Houston scrubbed a planned Station 10 and directed them back to the lunar module. They pulled up outside *Challenger* caked with dust. After they unloaded their samples and gear, Schmitt walked out to the ALSEP array for some final housekeeping, and Cernan steered the rover toward its final resting place.

He parked it on a slight rise 518 feet from the lander, giving the TV camera a nice, wide view of liftoff. He aligned the high-gain antenna, dusted the LCRU, and gave the batteries and console a final brushing. "Boy, I tell you, I ain't going to do much more dusting after I leave here," he told Parker. "*Ever.*" Off-camera, he pulled the substitute fender off the right rear wheel and the factory fender extensions off the others. They're in museums today.

"Okay," he said, aiming his camera at the rover. "Let me get one parting shot—one of the finest-running little machines I've ever had

LRV-3, stripped of its fender extensions, sits ready to capture the Apollo 17 crew's departure from the moon on December 14, 1972. (NASA)

the pleasure to drive." LRV-3 had covered more than twenty-two miles of the Taurus-Littrow's rugged lurrain. Cernan had spent four hours, twenty-six minutes driving it. And the machine had hauled 243 pounds of rocks and soil from the mission's far-flung sampling sites. On the walk back to the lunar module, Cernan thanked "the thousands of people in the aerospace industry" for Apollo's success, adding: "There's a lot that goes to getting this rover running out here that we don't have much to do with."

With Schmitt already in the lander, the last man and last driver on the moon ascended the ladder. On the climb he passed a stain-less steel plaque attached to the LM. "Here, man completed his first explorations of the moon," it read. "May the spirit of peace in which we came be reflected in the lives of all mankind." The moonwalkers of Apollo 17 spent the night in the LM. When they woke, Schmitt recited a poem he'd composed, a variation on "The Night Before Christmas" featuring "a miniature rover and eight tiny reindeer." A few hours later, with the command module in position overhead, *Challenger*'s ascent stage blasted free of the surface.

This time Ed Fendell got it all. He kept the craft on screen for twenty-six seconds, through its rapid ascent and pitchover to horizontal flight. The rover's TV camera kept running as the ship disappeared into the black.

PART SEVEN

TIRE
TRACKS

63

AS THE ASTRONAUTS OF APOLLO 17 FLEW HOME, EVERYONE AT NASA understood that American rockets would be sticking a lot closer to home in the coming years. In giving the green light to the Space Shuttle, the administration of President Richard Nixon had obligated the agency to orbital missions in place of epic voyages. Certainly there were valuable discoveries waiting in low Earth orbit, though driving the decision, above all else, was money. Frequent flights with a reusable spacecraft would better fit the agency's dwindling means and address one of the enduring complaints about the Mercury, Gemini, and Apollo programs: most of their hardware, billions of dollars' worth, had been used once and discarded.

But the assumption within the agency was that a return to the moon was just a matter of time—that the research already conducted into more lavish missions and longer-term stays would not, could not, go to waste. Surely future astronauts would crisscross the regolith in something resembling a Molab. An eventual base up there seemed inevitable.

No one believed that more than Wernher von Braun. Just before Apollo 17's launch, he predicted that in a few years' time, the Space Shuttle would be ferrying travelers to "space tugs" waiting in Earth orbit for trips to the lunar surface, "so for the entire flight to the moon and back, reusable equipment elements will be used." The space tugs would haul not only people but "very sizable payloads to the moon which may include temporary or semipermanent housing facilities.

"Then there will probably be surface vehicles with pressurized cabins on top that will have ranges of a thousand miles or so," he predicted. "So I think by 1990, you can definitely expect things like a traverse of the moon in pressurized surface vehicles. There will be at least temporary stays where people will stay a couple of weeks."

That didn't stray far from the vision he'd first shared with the public twenty years before. Von Braun had high hopes that Apollo 17 would open the world's eyes to the promise of, and the need for, return trips—that it would spark "the realization that we have only begun the exploration of the moon." And that, eventually, time spent there would serve as a springboard for the logical next step in space exploration: manned missions to the planets.

It didn't work out quite that way. Years after Apollo 17's return, however, the long drive to create the lunar rover did inspire a new cadre of scientists and engineers. While most of NASA was occupied with building and flying the Space Shuttle, the Jet Propulsion Laboratory held on to the six-wheeled, flexible Surveyor LRV model that Greg Bekker and Frank Pavlics designed and built in 1962. Eventually the lab examined the model anew, and found much that it liked for a mission von Braun had dreamed about since his days at Peenemunde: exploring Mars.

The Pasadena lab's people agreed that six wheels were far more capable than four. They were intrigued by the possibilities of its flexible frame. But a mechanical engineer there, Donald Bickler—who told me he conferred with Pavlics on GM's rover models back in the sixties, and with whom Pavlics now consulted on regular drives down to JPL—conjured a simpler, tougher variation on Santa Barbara's central idea. Working on his own time, in own his garage, Bickler replaced GM's flexible frame with a rigid but limber suspension system. He attached the three wheels on each side not to the vehicle's body but to a separate set of pivoting "rocker-bogie" struts that permitted the wheels to move independently. The system enabled the wheels to help one another over obstacles much as the GM model's had—they could climb stairs—while employing no elastic elements. Not so much as a spring.

Bickler patented the design in 1989, and, seven years later, the rocker-bogie suspension was a signature feature of Sojourner, a remote-controlled minirover shot into space as part of the lab's Mars Pathfinder project. Twenty-six inches long and weighing just twenty-four pounds, Sojourner was designed to pick its way across the rocky surface of a Martian floodplain, shooting pictures and chemically analyzing rock and soil samples. In all, it crept about 330 feet at a top speed of sixteen inches a minute. The robot trumped NASA's fondest hopes for it. Sojourner was designed to survive a week. It lasted three months.

Von Braun wasn't around to applaud this development. He had died of cancer, at sixty-five, five years after he left NASA. By then, his greatest achievement, the Saturn V, had made its final flight, unmanned, to lift the Skylab space station into orbit, and the Marshall Center was a very different place—purged of its remaining German leadership, none too gently, after Eberhard Rees's 1973 retirement, and reorganized to reflect NASA's shifting priorities.

Greg Bekker had died, too, after a long and busy retirement. Between gigs as a consultant and a speaker, he'd taken up watercolor painting, his favorite subject the Apollo rover; winning the LRV contract, he wrote, had been "a culminating point of my 40 years professional career." After his death, his native Poland recognized his lifetime contributions to transportation engineering, something it had not done during the Cold War. "We never forgot Professor Bekker," an admirer at the Warsaw University of Technology said, "but for years, we could only whisper his name in the hallways."

The rocker-bogie's success on the Pathfinder mission recommended it for the next generation of Mars rovers, Spirit and Opportunity, which landed on opposite sides of the planet in January 2004. Much larger, heavier, and stabler, they were built to undertake ninety-day quests for evidence that water once flowed there. Like Sojourner, they far outstripped their design specs: Spirit lasted six years. Opportunity lasted more than fourteen, during which it covered more than twenty-eight miles, giving it the off-Earth land distance record. Both found ample evidence that the Red Planet was once quite wet.

By now, the rocker-bogie was NASA's go-to rover design. In November 2011 it launched an even more ambitious Mars machine, a rolling laboratory called Curiosity. On the move at a top speed of eight feet a minute, the car-sized machine navigated and chronicled its travels with seventeen cameras. Its seven-foot robotic arm included a drill for collecting powdered rock samples, which it analyzed in an onboard lab. It fired a laser at rocks from a distance, then read the resulting vapors to identify their geologic makeup. It could climb obstacles at least as high as its twenty-inch wheels, which had aluminum rims over curved titanium spokes—a construction that recalled Bendix and Grumman designs of the 1960s.

Curiosity's nearly decade-long hunt for evidence that Mars once supported microbial life is still going as I write this. In February 2021, it was joined by an even more capable rover, Perseverance, which is scouring a dry lake bed for further signs of life, collecting rock and soil samples, and analyzing the planet's habitability for future human visits. It even carries a robotic helicopter to scout the surrounding countryside.

Sam Romano lived to see most of these amazing advances. General Motors transferred him to Detroit after the LRV project, and he later spent a few years working on fuel cells at Washington's Georgetown University, but after both stints he moved back to Santa Barbara. He died there in 2015 at eighty-seven. His widow, Marge, still has the wire-mesh rover wheel that his colleagues presented him when he left the GM lab in 1974. "May it serve as a reminder of 'the good old days,'" a plaque on the wheel reads, "and those who shared them with you."

64

FIFTY YEARS ON, THE ORIGINAL LRVs SIT JUST AS THE ASTRONAUTS parked them, frozen in time like unearthed children of Pompeii. A Bible still propped against the hand controller of the machine at Hadley Base. Dust thick on the floorboards of Rover 2 at Descartes. The third, stripped of its fender extensions, on a knoll overlooking the landing site at Taurus-Littrow.

The machines were among a mountain of gear left behind by the six Apollo missions to reach the moon. The ALSEP instruments remain in place, as do the lunar modules, their descent stages now and forever the centerpieces of the landing sites, their ascent stages crashed into the moon after delivering their crews back to the mother ships. The spent Saturn V third stages used on Apollos 13 through 17 were aimed at the lunar surface, too, and lie ruined in their own craters. And around each base camp lie scattered cameras, geologic tools, boots, spare gloves, gear boxes, empty pallets, and litter—food containers, unused sample bags, strips of insulating foil, and shredded Styrofoam packing, melted and shrunken, its air bubbles burst.

We can see a lot of it. Thanks to NASA's Goddard Space Flight Center in Greenbelt, Maryland, we have the lunar reconnaissance orbiter, a robotic spacecraft that has circled the moon since June 2009, poring over its surface in greater detail than ever before. The two-ton craft is packed with sensors and with three cameras—two with narrow-angle, telescopic lenses, and one taking a wide-angle view of the lurrain below.

It was a cinch that the cameras' principal investigator, Mark Robinson of Arizona State University, would direct the orbiter's eyes toward the Apollo landing sites. "The most common questions asked of me before the LRO launch related to seeing the astronaut tracks and the flags," he told me. The orbiter didn't disappoint. At a resolution of roughly twenty inches per pixel, its photos captured the descent stage of each lunar module, the ALSEP arrays, and the tracks of the astronauts between and around them. And at three of the sites, there were the rovers: well-defined rectangles, with their snouts aimed at the lunar modules and their wheels discernable at their corners.

I speak from experience when I say that it's easy to lose hours studying the pictures online, and that their most impressive details might be the tire tracks. They appear as crisp parallel lines spraying from the base camps and across the regolith. You can follow them out to the science stations: LRV-1's tracks are visible on the traverse to the Hadley Rille, the climb up Hadley Delta to Station 6, and around the green boulder where Jim Irwin had to hold the rover to keep it from sliding downhill.

From above the Descartes Highlands, the orbiter's pictures captured the Apollo 16 crew's travels distinctly enough that Charlie Duke, in describing how well John Young controlled LRV-2, was able to tell me: "You look at our tracks that we left behind, and they run pretty straight." And at Taurus-Littrow, rover tracks and footprints surround Ballet Crater and the crew's other stops. You can even make out some of the small craters that Gene Cernan chose to drive straight through.

"To this day, seeing the tracks is somewhat of an inspiring experience," Robinson said. "We used to go to the moon!"

65

SONNY MOREA AND FRANK PAVLICS ARE SOMETIMES ASKED whether a returning astronaut could climb into one of the Apollo rovers and fire it up. Morea told me he reckons a half century of exposure to the lunar elements—the constant bombardment of cosmic radiation and micrometeoroids, and the regular cycle of temperature extremes—have left the machines brittle and fried. Metal parts that once spun have likely seized up, thanks to the moon's hard vacuum. Wiring is shot. The wire-mesh wheels might crumble under load.

Pavlics figured that, at a minimum, they'd need work. "They probably would need fresh batteries and the control electronics," he said. "I'm sure that after going through so many heat and cold cycles the [circuit] boards are cracked, and all that. So you would need a new control box, and, of course, new batteries.

"There *is* a problem with what we called cold welding, with metal-to-metal contacts: with longtime exposure to vacuum, they are kind of frozen together. So, they'd have to be broken apart, forced to move, so there could be a sliding motion between the metals. But besides that, the mechanical parts should be working, I think."

I asked Pavlics whether he would build a different rover today, what with all the advances in technology since 1969. "The basic rover would be the same," he replied. "The size and configuration were pretty much dictated by the landing machine—the lunar lander. That probably couldn't have changed much." He might trade the rover's parts for new composites and alloys, however: "We did use titanium quite a bit, but nowadays there are better materials, so we could have made it even lighter in weight."

Otha "Skeet" Vaughn, a Marshall Center engineer whose study of the moon's soil was used to write the rover's performance requirements, was keen on the original design when I asked him the same thing. "If we went to the moon today, you could go with the same vehicle—don't change a thing about it," he said. "It was a successful vehicle. It traveled *fifty-six miles.*"

Indeed, it did. All of the J mission improvements were important— the beefier lunar module enabling an extra day on the surface, improved backpacks and space suits that extended each extravehicular activity, refined geologic tools. But none so fully transformed the missions as the rovers. With Apollo 15 came exploration measured in miles, not minutes.

Were the machines worth their expense, not to mention the heartache and stress their hell-for-leather creation inspired? No argument, they didn't come cheap. Including the price of development, each of those fifty-six miles cost something in the neighborhood of $680,000, or well over $4 million in today's money. And that doesn't include what GM spent out of its own pocket before the "We must do this!" meeting, or the millions Bendix and Grumman invested in their ideas, or the many contracts throughout the 1960s for earlier moon vehicle concepts that informed the final product.

Yet the rover's price, however dear, looks like a bargain in the context of the overall Apollo program, which cost upward of $25 billion in the dollars of the time. NASA pegged the cost of Apollo 15 alone at $445 million in 1971. LRV-1 accounted for less than 3 percent of that, tops. And as Rocco Petrone said, "it made the mission."

As for the breathless pace of the rover's creation, well, it took a toll. Pavlics lamented that "the kids didn't see me much." Gene Cowart recalled that the job's pressure made some of his colleagues sick and left its marks on him. "I'm thirty-nine, but I worked on this thing," the Boeing engineer told me in 2019; in truth, he was born in October 1923. "I wouldn't want to undertake it again. I think I'd rather reenlist." Morea remembered the experience as distinctly unpleasant, with few days that weren't anxious and difficult. "I marvel that we were able to accomplish what we did, knowing what I know now," he said. "I marvel that they didn't replace me as project manager."

Shortly before Christmas 1970, as it became clear that the first rover just might make its deadline for Apollo 15, Morea's bosses asked him to share the lessons he'd take away from the affair. His twelve-page answer suggested changes to NASA's procurement rules, the makeup of source evaluation boards, the criteria for selecting contractors, the style of the government's contracts with them, and the formulas it used to project costs. He also singled out the "difficult position [Marshall] assumed in designing a spacecraft type of machine which another Center's astronauts will use, which flies [sic] as part of another Center's spacecraft and which was openly opposed by many segments of that other Center for the first six months of the program." He closed his remarks under the headline: "Let's not ever take on a task like LRV again!"

Still, its challenges notwithstanding, Morea recognized the rover as a thrilling addition to the space program—and, yes, worth the anguish. "Otherwise the astronauts would just have kept bouncing around in their space suits, which they couldn't do for long," he said. "The LRV was not a huge technical challenge. It was a timetable challenge. Every day, something would come up that threatened to make the schedule slip.

"But you had a lot of people on this thing, all trying to get it done in seventeen months. And it all came together."

Greg Bekker came to a similar conclusion. While he lamented the job's "burocracy" [*sic*] and the feeling among his GM colleagues that it "was exhousting [*sic*] and, perhaps, unrewarding for those who carried the burden of making every bolt and nut work," he recognized it as "part of such a gigantic undertaking that personal frustrations, which are inevitable in any business, have no meaning."

In retrospect, the frenzied nature of the rover's creation, and the hassles that NASA and its partners ran up against during those hectic months, serve only to underline the remarkable nature of the achievement. Cowart offered one measure of how successful the project is seen as today: "You go around Huntsville, and everybody you meet will tell you he worked on the rover," he said. "You can't find anybody who *didn't* work on this thing."

Those actually responsible tend to be modest about it. Up on a wall in Frank Pavlics's home office, hanging beside the wire-mesh wheel, are framed black-and-white portraits of space-suited astronauts. "To Ferenc Pavlics," reads an inscription over one fading image, "with best wishes from the crew of Apollo 15 and thanks for your efforts which contributed to the success of 'Rover One'." Outside that room, his home offers no hints of his contributions. His one flourish of self-promotion was intended for very few eyes: the undercarriages of all three rovers, he told me, bear his signature.

Morea, too, prefers to let the work speak for itself. He spent another eighteen years at the Marshall Center once he'd wrapped up the rover project, serving as assistant director of the Structures and Propulsion Lab, then as the center's assistant director for research and technology. He and Angela had two more children; two of his daughters grew up to be engineers.

He chose to stay close to his life's work after retiring in November

1990—in a home on Huntsville's south side, a ten-minute drive from the U.S. Space and Rocket Center. He volunteered every Wednesday there for years, arriving in the morning and visiting with schoolchildren and tourists, most often next to the LRV at the Davidson Center's west end. After a break for lunch amid the noisy fuss of grade school field trips in the center's Mars Grill (home of the Astro Burrito and corn dogs trading as Pluto Pups), he'd return to the Davidson for hours in the afternoon.

The virus put a halt to that, shutting down the Space and Rocket Center. By the late summer of 2020, the museum was down to a skeleton staff and publicly appealing for donations from the alumni of its Space Camp program to get through the fall.

66

NASA HAS NO ANNOUNCED PLAN TO SEND ANYONE OR ANYTHING back to the lunar regions explored on the Apollo missions. But in a development that would gratify Wernher von Braun and everyone else involved with those long-ago visits, it *is* preparing to dispatch a rover to never-explored lurrain up there, and soon. In February 2020 the agency contracted with a Pittsburgh company, Astrobotic, to fly its VIPER rover to the lunar south pole in 2023.

The brainchild of NASA's Ames Research Center in Silicon Valley, VIPER—another agency acronym, short for "volatiles investigating polar exploration rover"—is set to explore the permanently shadowed craters and valleys there for water ice, as a first step toward harvesting the moon's resources to support a base. That the rover will find ice is not in doubt, assuming the machine survives the trip through space; the questions are how much exists and how accessible it is from the surface. If there's a lot, and it's easily mined, it could theoretically be used to make air, water, and fuel, eliminating the need to haul them from Earth. So the machine will descend to the pole aboard a privately designed and built lander that in many respects evokes the Soviet Luna 17 that deposited Lunokhod-1 on the moon in 1970. The Astrobotic spacecraft will deploy ramps, just as the Luna did, and VIPER will roll off and away.

The polar mission poses new challenges. The sun hangs far lower in the sky, even at lunar midday, than it did at the Apollo sites. Navigation in the gloaming will be difficult. Capturing the sun's energy via VIPER's vertical solar panels will be more sporadic. Surviving the pole's unimaginably cold permanent shadows, where the rover will prospect for hours at a time, will be a job in itself. The machine chosen for this demanding environment likely would alarm Greg Bekker and Frank Pavlics: it's a blocky, half-ton craft on four smallish wheels that, from a distance, doesn't look especially rough and ready—imagine Lunokhod crossed with Bendix's ill-fated Surveyor LRV model. But it is far more agile than appearances suggest. "It looks like the center of gravity would be really high because it's a big, boxy thing," William Bluethmann, NASA's VIPER team leader at the Johnson Space Center, acknowledged. "But we keep the batteries"—by far the heaviest components—"down low." The wheelbase, a little more than a yard square, gives it a wide stance that enables the rover to tilt more than forty-five degrees without losing its balance. Its seemingly too-small wheels can each independently turn, so that it can sidestep or spin in place if need be.

And its suspension is intriguing. The VIPER team chose an active design that enables controllers to independently adjust the ride height over each wheel—or to lift a wheel off the ground to "step" over or around hazards. In the loosely compacted soils NASA expects near the landing site, that could prove essential to avoiding a fate worthy of Thomas Gold. "You lift a wheel up, a wheel that's stuck in soft soil— you lift it up out of it, and you move it to firmer soil," Bluethmann said. The suspension also enables VIPER to hike itself up and over rocks. On steep climbs, "you can lean the body uphill."

Bluethmann and his team spent months studying past mobility

ideas, including those of Bekker and Pavlics, before deciding on its approach. "We did a deep design on this," he told me. "What we were doing was probably not that dissimilar to what they were doing in the sixties." Often, he said, "we'll think we have a great idea, and look back at what those guys did, and see that they had a lot of the same ideas." Though its four drive motors are mated to traditional planetary transmissions, the smaller motors steering the wheels and adjusting the suspension use harmonic drives.

I hesitate to mention VIPER, inasmuch as the project is the latest in a long procession of lunar mobility schemes, all of which have so far failed to pan out. VIPER is directly descended from Resource Prospector, a similar rover that NASA snuffed in 2018. Before that, the agency touted its space exploration vehicle, a modular cabin that could be flown around in space or mounted on a six-wheeled chassis for travel on the lunar surface. The ground version amounted to a reimagined Molab and was intended for use by 2020. The sidelined projects go on and on, back through decades. When it comes to returning to the moon, NASA has accomplished more dreaming than doing.

Which brings to mind a conversation that took place shortly before the last of the Apollo rovers was parked and its driver walked away. Apollo 17's Gene Cernan and Jack Schmitt were relaxing in the lunar module after their record-setting second EVA, preparing for sleep and chatting with capcom Joe Allen. "There are a lot of us looking forward to that third EVA tomorrow," Allen told them. "It's going to be the last one on the lunar surface for some time."

Everyone on the ground had marveled at the color and sharpness of the rover's TV pictures, and Mission Control had enjoyed seeing "the tracks that you're leaving behind in the lunar soil, both footprints and rover tracks," he said. "And some of us are down here now, reflecting on what sort of mark or track will—someday—disturb the tracks that you leave behind there tomorrow."

Cernan and Schmitt considered the comment. "That's an interesting thought, Joe," Cernan told him. "But I think we all know that somewhere, someday, someone *will* be here to disturb those tracks."

"No doubt about it, Geno," Allen said.

Schmitt must have detected doubt in the capcom's voice. He told him to buck up, adding: "I think it's going to happen."

"Oh, there's no doubt about that," Allen said, "but it's fun to think about what sort of device will ultimately disturb your tracks."

"Well," Schmitt replied, "that device may look something like your little boy."

Allen's son, David, was four years old at the time.

He's in his fifties now.

Neither he nor anyone else has been back to mess up those tracks.

ACKNOWLEDGMENTS

The idea seemed so straightforward, so alluringly simple, when my editor broached it in a February 2019 email: "I'm fascinated by the 'moon cars' that were driven during Apollo 16 and 17, the last lunar missions," Peter Hubbard wrote. "I have a nutty idea that a very short but deeply reported book on those 'road trips' could be cool."

I have enough techno geek in my genes that the subject immediately appealed to me. But as I was soon to find out, nothing about it was simple: any story involving the space program opens a spray of wormholes twisting ever deeper into the arcane. To write about this tiny sliver of Apollo required grounding in a host of unexpected topics—and the more I learned, the more remained for me to know.

So I owe a lot of people for smartening me up. First and foremost is Sonny Morea, who has been generous with his time and knowledge on my trips to Huntsville and in our many phone conversations. He discussed the rover project's pitfalls as forthrightly as he did its triumphs, and guided me through NASA's nuanced 1960s culture toward something that pretty closely resembles an understanding.

Ferenc "Frank" Pavlics was likewise generous with his memories and patient with my questions, both during my visits and in numerous follow-up contacts. His son Peter proved a willing and genial go-between when Frank's health made that necessary. I thank both of them.

Others who built the rover gave real texture to the story. Among the alumni of the Marshall Space Flight Center, I thank William W. "Bill" Vaughan, Ron Creel, Otha "Skeet" Vaughn, Kenneth M. "Mike" Grant, and Craig Sumner (who actually still works there). From among those who worked at Boeing Huntsville, I thank Eugene Cowart and Al Haraway. I'm grateful to three veterans of GM's Santa Barbara operation: Paul Jaquish, John Calandro, and Jerry Skaug. And I benefited from conversations with former Bendix staffers Bernard Pilon and Sam Fine.

As you'd expect, astronauts provided key ingredients. Dave Scott shared helpful documents and offered feedback on my chapters on Apollo 15. Jerry Carr treated me to a witty and wry account of his role in developing the LRV. And Charlie Duke left his fingerprints on various parts of this tale: he spent hours on the phone with me, read over my chapters about Apollo 16, and gamely answered my emailed questions. Thank you, General.

At the Johnson Space Center, William Bluethmann walked me through the upcoming VIPER mission, and Don Bickler, late of the Jet Propulsion Laboratory, offered an entertaining shotgun blast of recollections about the rocker-bogie. I thank them both. I relied several times on the skills and hustle of NASA public affairs officers, and really appreciate their efforts on my behalf: Brandi K. Dean of the Astronaut Office at JSC, Andrew Good at JPL, and Alison Hawkes at Ames Research Center.

My favorite few days of this project might have been those I spent in Flagstaff, and in particular the afternoon that Laszlo Kestay took me out to the Cinder Lake crater fields. I couldn't have asked for a smarter, more upbeat host. Retired geologist Jerry Schaber had me over to his house, and the insights I gleaned on that visit inform much of the rock science in this story. Kevin Schindler, the historian at the Lowell Observatory, helped put me in touch with both of these men and provided me with context for the Geological Survey's contributions. The pandemic upended my attempts to interview Putty Mills in person, but I thank his daughter, Denice Dogan, for serving as a bridge between us.

My old friend Steve Beschloss arranged for my visit to Arizona State University, where I was fortunate to spend a little time at its School of Earth and Space Exploration—home base of Mark Robinson, the principal investigator of the lunar reconnaissance orbiter cameras. I thank Mark for suffering my questions about that amazing program.

Some of the story's principals are no longer with us, but with the help of their families, I was able, I hope, to do them justice. Marge Romano and Cliff Romano were great helps. Tim Romano went above and beyond—he and I have corresponded by email for months on details of his father's life and work, and he trusted me with Sam Romano's photos and LRV project diaries, which were fascinating windows into the man and the rover's difficult birth.

Likewise, I have the family of Edward G. Markow to thank for my depiction of the Grumman engineer: his wife, Margaret; son, Jim; and daughter, Elizabeth Markow-Brown. I also thank Georg von Tiesenhausen Jr., whose father produced an early rover design while on the von Braun team at ABMA.

Late in this project, I was fortunate to make contact with Professor Andrzej Selenta of the Warsaw University of Technology, who gave me the information I needed to track down Greg Bekker's daughter, Dr. Eva Heuser, and his granddaughter, Joy Otsuki. All three of these generous people proved key to my passages on Bekker's life—though, regrettably, I had to leave a lot of detail out of the finished text. For instance: Bekker was a virtuosic whistler, with a repertoire ranging from bird calls to concertos. Who would have guessed?

This story relied on a mountain of documentation, which I was able to find only with expert help. I thank Drew Adan, Amy Wells, and Caroline Gibbons at the University of Alabama in Huntsville's Salmon Library; Shane Bell and his colleagues at the National Archives in Atlanta; Rodney Krajca at the National Archives in Fort

Worth, Texas; Audrey Glasgow and Pat Ammons at the U.S. Space and Rocket Center; and Daisy Muralles at the University of California, Santa Barbara. Tom Cattoi of FLIR Systems, who oversees the property once home to GM's Santa Barbara operation, kindly dug up and shared photos of the place as it looked in its rover days.

I'd have missed out on a wealth of material without help from the folks at SAE International, in the past known as the Society of Automotive Engineers, who gave me access to documents preserved on the society's online portal, SAE Mobilus—and, near as I can tell, nowhere else. Thank you, Mike Paras, Kimberly Martin, and Erin Renoll.

Thanks, too, to Larry Kinsell of General Motors, who made available photos of the company's many rover concepts; Nan Halperin of the Science Photo Library; the aforementioned Bernard Pilon, who entrusted his cache of rare Bendix rover photos to my care; and my brother Kevin, who created the handsome maps of the LRVs' lunar traverses. I owe particular thanks to Don McMillan, who produced the beautiful diagram of the rover in the photo insert.

This book also owes much to two experts who allowed me to steal their time and borrow their big brains. Mike Neufeld is a wellspring for anyone writing about Wernher von Braun. His biography—*Von Braun: Dreamer of Space, Engineer of War*—is the best book about the enigmatic German, hands down, and I reckon it's destined to remain so. This generous man took pains to ensure I had any useful information he could offer. I'm deeply grateful to him.

Eric M. Jones, creator of the online Apollo Lunar Surface Journal, spent years assembling what is, in my view, the most useful resource on the moon landings—a word-for-word transcript of everything said on the surface, annotated with photos, video and sound clips, related documents, and insights from ten of the twelve moonwalkers. He read over all the chapters in Part Six and answered my questions on Apollo esoterica.

Across the Airless Wilds is the third book on which I've worked with Peter Hubbard, and I'm reminded that I have an editor most writers would kill for, if they knew what they were missing. As always, he's been a smart, funny, encouraging presence throughout the process, a thoughtful collaborator, and a tireless champion of the results. I owe him even more this time around, because as I mentioned, he came up with the idea.

My agent, David Black, is the best there is at what he does. We've worked together for nearly twenty years, during which I've never stopped wondering how I could have so lucked out. He is fierce. He is brave. He is a warrior in the cause of righteousness—and me.

Finally, I'm much obliged to those people who kept me moored to reality while I was chained to my keyboard, a period that coincided with the emergence of Covid-19 and the strange and terrible times that followed. Amy Walton has backed this project from its start, and supported me for nearly two decades. I'm deeply thankful to have her in my life.

My dear friend Laura LaFay has read every word between these covers, several times. She's one of the best writers I know, and an incisive reader and editor. I don't write anything these days without asking her to give it a look, and though her advice is sometimes painful to hear, I almost always follow it. More than that, she's a stand-up human being. Thank you, Magua.

Mark Mobley, my friend for more than thirty years, read the manuscript and offered good ideas about how to strengthen it. But more importantly, he and I have talked pretty much every day since I started work on it—and his friendship has been among my essential tethers to normalcy.

All the while, my daughter, Saylor, has graced me with an abundance of love and laughter. She's the only human I've spent regular time with in more than ten months, as I write this, and in each of our socially distanced contacts, she's given me a new reason to be proud of her. If you wanted the moon, Sweetie, I'd do my damndest to get it for you.

NOTES

This is a work of nonfiction. As such, it relies on interviews with participants in the events I describe, backed up and augmented by documentary evidence. All of the interviews are listed below. Most of the documents are, too, along with where you'll find them.

Be advised, however, that this is intended as a popular history rather than an academic treatise. So while I identify the collections housing the documents, you'll not find box and file numbers listed. All but one of these collections are fairly small, and the exception—the ninety-seven boxes of material at the National Archives in Atlanta—are so labeled that chasing down a specific piece of paper doesn't require much in the way of intuition.

Why have I done this? Because (a), these notes are already so outrageously long that they wouldn't fit in the book if I included that stuff, and (b), the notes are meant to be *read*—there's interesting background and detail here, and asides that didn't fit comfortably into the narrative, and a little informed speculation, and links to valuable online sources that are worth your time—and I find that nothing discourages reading quite like signaling up front it's going to feel a lot like work. So.

As a general rule, when the personal recollection of an interview subject, decades after the events in question, contradicts information committed to paper while or immediately after the events occurred, I gave greater weight to the documents. I made one or two exceptions to that rule, which I note.

The collections are abbreviated, as follows:

NARA-Atlanta—the National Archives at Atlanta, the repository for most records from the Marshall Space Flight Center. Unless otherwise noted, these records include all of the center's "weekly notes" listed below, as well as Sonny Morea's inputs to them, which he usually put together a day or two ahead of time; all of Boeing's monthly progress reports on the LRV; the Boeing contract; and the Boeing proposal. In other words, if you see a reference to "weekly notes" or the Boeing contract below, you'll not find collections listed for them, because I just told you they're from NARA-Atlanta.

NARA-FW—the National Archives at Fort Worth, the repository for most records from the Manned Spacecraft Center in Houston, today called the Johnson Space Center. Those pertaining to this story are mostly about astronaut training and geology.

UAH—the LRV and Morea collections at the Salmon Library, University of Alabama in Huntsville. These include handwritten notes Sonny Morea kept during the project.

UCSB—the Delco Retirees Group collection, among the special collections at the University of California, Santa Barbara.

PART ONE: THE DIFFERENCE IT MADE

1

The visit I describe took place on April 24, 2019.

The Davidson Center's dimensions are from the website of Metl-Span, the company that produced the building's wall panels, at https://www.metlspan .com/davidson-center-for-space-exploration/.

The Saturn V is described in a slew of NASA publications and websites. I'd start with the *Saturn V Flight Manual,* prepared for Apollo 8 in the fall of 1968 and available on the NASA History section of the agency's website at https:// history.nasa.gov/afj/ap08fj/pdf/sa503-flightmanual.pdf. See also W. David Woods, *How Apollo Flew to the Moon* (New York: Springer, 2011), a terrific summary of the hardware, written by the creator of the valuable online Apollo Flight Journal, and Boeing's historical snapshot of the rocket, on its website at https://www.boeing.com/history/products/saturn-v-moon-rocket.page.

2

My passages on the Apollo 14 and Apollo 16 traverses relied on the Apollo Lunar Surface Journal, a deeply researched online resource that offers transcripts of, and background on, everything said and done by astronauts on the moon. Assembled by Eric M. Jones, who was assisted by a large supporting cast (including ten of the twelve moonwalkers), the ALSJ stands today—and, I hope, forever—as a priceless public service. You can find its main page at https://www.hq.nasa.gov/alsj/main.html. The start of the Apollo 14 hike is preserved at https://www.hq.nasa.gov/alsj/a14/a14.staA.html, and the drama continues at https://www.hq.nasa.gov/alsj/a14/a14.tocone.html. The account of Apollo 15's journey is from the chapter titled "The First EVA" at https:// www.hq.nasa.gov/alsj/a15/a15.html.

For all of the book's moon scenes, I checked the ALSJ against the official mission transcripts, preserved as PDFs on the Johnson Space Center's History Portal website. The Apollo 14 EVA can be found at https://historycollection .jsc.nasa.gov/JSCHistoryPortal/history/mission_trans/AS14_LM.PDF (in the section beginning on page 4.1). The Apollo 15 transcript, organized differently, is at https://historycollection.jsc.nasa.gov/JSCHistoryPortal/history /mission_trans/AS15_TEC.PDF; the first EVA begins on page 437 (all accessed August 21, 2020).

3

NASA tweeted a map of the Apollo 11 crew's wanderings, overlaid on a football field, in February 2019; you can find it at https://twitter.com /NASAMoon/status/1092239824411639809/photo/1. Why it didn't tilt the traverse map to fit within the sidelines is a mystery, but you can see that it would fit just fine. The agency overlaid other Apollo traverses over features on Earth, to great effect, at https://www.hq.nasa.gov/alsj/TraverseMapsEarth .html (accessed September 11, 2020).

The lunar reconnaissance orbiter camera's main site is at LROC online, https://www.lroc.asu.edu/.

PART TWO: NATION OF IMMIGRANTS

4

My description of von Braun's early life was informed by Michael J. Neufeld, *Von Braun: Dreamer of Space, Engineer of War* (New York: Knopf, 2007). Von Braun described getting his first telescope in "Space: Reach for the Stars," *Time,* February 17, 1958. His famous "Columbus" quote is from Daniel Lang, "A Romantic Urge," *New Yorker,* April 21, 1951, as is his "We needed money" quote.

Von Braun's wartime career is detailed in Michael J. Neufeld, "Wernher von Braun, the SS, and Concentration Camp Labor: Questions of Moral, Political, and Criminal Responsibility," *German Studies Review* 25, no. 1 (February 2002). See also Neufeld, *Von Braun;* and Wayne Biddle, *The Dark Side of the Moon: Wernher von Braun, the Third Reich, and the Space Race* (New York: W. W. Norton, 2009).

One needn't study von Braun's biography in depth to feel deeply ambivalent about the man. That he was willing to apply his embarrassment of talents to the Nazi war effort is troubling on its face. That he did so, as he later insisted, in the name of pursuing his personal goal of spaceflight, comes off as a Faustian bargain that contributed to thousands of deaths. Neufeld, a former chair of the Space History Division at the Smithsonian National Air and Space Museum, told me he continues to wrestle with "the complexity of the problem: on the one hand, there's no doubt about the brilliance of his engineering leadership in Huntsville. How do you balance that against his actions at Peenemunde? I'm constantly steering between these poles, between his technological accomplishments and the war criminal."

The "I started out" quote is from a speech von Braun gave to the Sixteenth National Conference on the Administration of Research, in French Lick, Indiana, on September 18, 1962. The Marshall Center's Bill Vaughan gave me a copy of it in June 2019; conversations with other retired Marshall engineers revealed that they beheld the speech with a reverence typically reserved for scripture. You can find it at Tim Reyes, "Retrospective: A Speech by Wernher von Braun on Management," Medium, last modified September 22, 2015, https://medium.com/@telluric/dr-wernher-von-braun-director-96eeae675528. One note: the speech is usually said to have occurred at a conference on "the *Management* of Research," and the linked transcript is so marked. The conference's proceedings use "Administration," as does press coverage of the speech.

Von Braun's SS membership was covered up by the federal government until after his death, when the details of Project Paperclip went public. His FBI file, in which it is mentioned several times, is available in the FBI Records: The Vault section of the bureau's website at https://vault.fbi.gov/Wernher%20VonBraun. It makes interesting reading.

General Dornberger's quote is from Walter Dornberger, *V-2* (New York: Bantam Books, 1979), available through the Internet Archive at https://archive.org/details/v2thebantamwarbo00walt/page/16/mode/2up (accessed July 29, 2020).

My passage on the Staveley Road V-2 attack was informed by articles on the Brentford & Chiswick Local History Society website: "Commemorating the Chiswick V2," https://brentfordandchiswicklhs.org.uk/local-history/war/commemorating-the-chiswick-v2/, and "The Chiswick V2," https://brentfordandchiswicklhs.org.uk/the-chiswick-v2/; Clare Heal, "The Day Hitler's Silent Killer Came Falling on Chiswick," *Sunday Express (U.K.)* online, September 7, 2004, https://www.express.co.uk/news/uk/508065/The-day-Hitler-s-silent-killer-came-falling-on-Chiswick; and by a short film, *First V2 Landing & Chiswick in the Blitz*, YouTube, 5:08, The Chiswick Calendar, https://www.youtube.com/watch?v=vnXaZJikGZ0 (all accessed July 8, 2020).

I also relied on the V2rocket.com website, a thorough resource on the weapon and its deployment. Its most impressive feature might be an attempt to account for every firing (and that event's effects) from September 8, 1944, to March 27, 1945, at http://www.v2rocket.com/start/deployment/timeline.html. General background on the rocket awaits at http://www.v2rocket.com/start/makeup/design.html (both accessed July 8, 2020).

Von Braun's future boss, U.S. Army major general John B. Medaris, heard the telltale double boom up close when his headquarters in France came under attack in September 1944: "I was startled by a heavy blast *followed* by a deep rumble that sounded like thunder," he wrote in *Countdown for Decision* (New York: Putnam, 1960). "I had had my first direct contact with Wernher von Braun and his team of liquid rocket experts."

My account of the V-2's scattered British toll was informed by the V2rocket.com timeline. For more on the Woolworth blast, see the Woolworths Museum website at http://www.woolworthsmuseum.co.uk/1940s-remembernewcross.htm. The Hughes Mansions attack also draws from V2rocket.com, and from Lianne Kolirin, "The Seder Bombing That Killed My Mother," *Jewish Chronicle* online, last modified May 10, 2016, https://www.thejc.com/news/features/the-seder-bombing-that-killed-my-mother-1.498532 (all accessed August 31, 2020).

Antwerp's experience under barrage, and the Rex Cinema disaster in particular, are described in "Antwerp—City of Sudden Death," a V2rocket.com article, http://www.v2rocket.com/start/chapters/antwerp.html (accessed August 31, 2020).

Von Braun's "in the hope" quote is from his French Lick speech.

The description of von Braun at his surrender is from a May 19, 1951, letter to the *New Yorker* from Bill O'Hallaren, former public relations sergeant of the U.S. Army's Forty-Fourth Infantry Division, and published in the magazine's May 26, 1951, edition.

Von Braun's predictive report is quoted from Neufeld, *Von Braun*.

5

The von Braun quote that opens the chapter is from his French Lick speech. In my copy, *spaceflight* is typed as two words; as von Braun calls it "a word bordering on the ridiculous," I've treated it as I believe he intended. The second von Braun quote is ibid.

The White Sands experiments are described in Daniel Land, "White Sands," *New Yorker*, July 24, 1948. *The Mars Project* is described in Neufeld, *Von Braun*.

Neufeld stressed to me that von Braun and his Germans were only peripherally involved in the Bumper launch, as it was overseen by a team from General Electric encamped at White Sands. Still, it was built on their shoulders.

The Dannenberg quote is from Shaila Dewan, "Town Transformed by Rocket Science Is Sure of Its Heroes," *New York Times*, January 3, 2008.

My description of 1950s Huntsville was informed by Luther J. Carter, "Huntsville: Alabama Cotton Town Takes Off into the Space Age," *Science* 155, no. 3767 (March 10, 1967).

The *New Yorker* profile was Lang, "A Romantic Urge," April 14, 1951.

Von Braun's fortuitous meeting with Ryan is described by Neufeld in *Von Braun* and by the same author in "'Space Superiority': Wernher von Braun's Campaign for a Nuclear-Armed Space Station, 1946–1956," *Space Policy* 22 (2006). See also Colin Davey, "San Antonio & the Genesis of the *Collier's* Series, 'Man Will Conquer Space Soon!' in *Horizons*, the online newsletter of the American Institute of Aeronautics and Astronautics' Houston Section, April 2013. It's available at http://www.aiaahouston.org/wp-content/uploads/2012/07/Horizons_2013_03_and_04_page_54_and_55_Colliers_series_Colin_Davey2.pdf.

Fred Whipple's quote is from his "Recollections of Pre-*Sputnik* Days," an essay in Frederick I. Ordway III and Randy Liebermann, eds., *Blueprint for Space: Science Fiction to Science Fact* (Washington, D.C.: Smithsonian Institution Press, 1992).

The AIAA Houston Section reprinted the *Collier's* series in *Horizons* over several months in 2012 and 2013. The first installment can be found at http://www.aiaahouston.org/Horizons/Horizons_2012_07_and_08_high_resolution.pdf (accessed August 14, 2020). It's a high-res scan of the entire August 2012 newsletter; the relevant material starts on page 42.

6

The second installment of the *Collier's* series is included in the September/October 2012 issue of the AIAA Houston Section's *Horizons* at http://www.aiaahouston.org/Horizons/Horizons_2012_09_and_10_high_resolution.pdf.

The third installment, from the November/December 2012 issue, is at http://www.aiaahouston.org/Horizons/Horizons_2012_11_and_12_high_resolution.pdf.

Regarding my reference to hydrogen peroxide: the V-2's engine ran on watered-down alcohol and liquid oxygen; H_2O_2 drove the turbine that pushed the fuel and oxidizer into the engine's combustion chamber.

The single exception I mention was Leon "Lee" Silver, a Caltech scientist who was allowed to speak briefly with Apollo 15's Dave Scott after his departure from the moon.

My passage on von Braun's Disney adventure was informed by Catherine L. Newell, "The Strange Case of Dr. von Braun and Mr. Disney: Frontierland, Tomorrowland, and America's Final Frontier," *Journal of Religion and Popular Culture* 25, no. 3 (Fall 2013); Randy Liebermann, "The *Collier's* and Disney Series," another essay in Ordway and Liebermann, *Blueprint for Space;* Neufeld, *Von Braun;* and repeated viewings of the *Disneyland* episodes on YouTube.

For more on the navy's Vanguard program getting the nod for the satellite launch, see Michael J. Neufeld, "Orbiter, Overflight, and the First Satellite: New Light on the Vanguard Decision," in Roger D. Launius, John M. Logsdon, and Robert W. Smith, eds., *Reconsidering Sputnik: Forty Years Since the Soviet Satellite* (Amsterdam: Harwood Academic Publishers, 2000).

7

My account of Bekker's early life draws from telephone interviews of January 29, 2021, with Bekker's daughter, Dr. Eva Heuser, and January 28, 2021, with his granddaughter, Joy Otsuki. The best public source is "The Polish Man Who Sent a Car to the Moon," on the Warsaw University of Technology website at https://www.pw.edu.pl/engpw/Research/Business-Innovations-Technology-BIT-of-WUT/The-Polish-Man-Who-Sent-a-Car-to-the-Moon (accessed September 1, 2020).

My reference to U.S. paved road mileage is from the Bureau of Transportation Statistics at https://www.bts.gov/content/public-road-and-street-mileage-united-states-type-surfacea (accessed July 8, 2020).

Bekker's study of the caterpillar track's history is detailed in his *Theory of Land Locomotion: The Mechanics of Vehicle Mobility* (Ann Arbor: University of Michigan Press, 1956). The following section is ibid.

My discussion of Bekker's movements during and after World War II draw from my January 2021 Heuser and Otsuki interviews. See also Andrew

Hills, "Modified Track Jeep, Online Tank Museum, last modified November 5, 2019, https://tanks-encyclopedia.com/coldwar-canada-modified-tracked-jeep/, and an article by the National Technology Museum in Warsaw, "Mieczyslaw Bekker (1905–1989)—Fly Me to the Moon," which I found at https://poland.pl/science/famous-scientists/mieczyslaw-bekker-19051989-fly-me-moon/ (accessed September 1, 2020). See also Joseph A. Reaves, "Poland Now Honors Lunar Rover Designer," *Chicago Tribune*, October 18, 1991, 22.

Bekker provides a nice summary of the slow development of his field in "Evolution of Approach to Off-Road Locomotion." *Journal of Terramechanics* 4, no. 1 (1967). His analysis of animal evolution is from his "Land Locomotion on the Surface of Planets," *ARS Journal* 32, no. 11 (November 1962), 1651–1659.

I found Bekker's Cobra work detailed on a blog (rather vaguely titled "For the Record") devoted to armored vehicles and video games involving tanks. Improbable though that might seem, the posting was thoughtful and thorough. The first of its three parts was at http://ftr.wot-news.com/2014/11/25/project-cobra-part-1-background-automotive-and-personnel-carriers/; the second at http://ftr.wot-news.com/2014/11/26/project-cobra-part-2-weapons-platforms-and-canadian-projects/; and the third at http://ftr.wot-news.com/2014/11/30/project-cobra-part-3-weapons-and-wot/ (all accessed November 18, 2020).

The quote regarding an engineer shortage is from Bekker's closing remarks to an April 1955 Interservice Vehicle Mobility Symposium at the Stevens Institute of Technology, available at https://apps.dtic.mil/dtic/tr/fulltext/u2/103663.pdf (accessed January 25, 2020).

8

I visited the Pavlics home on June 11–12, 2019. The entire chapter draws from my interviews on those visits, except as noted below.

The quote about his errant chemistry experiment is from David Clow, "The Law of the Stronger: Ferenc Pavlics and the Lunar Rover," *Quest* 18, no. 1 (2011).

I found the dates and length of the couple's transatlantic crossing in the ship's histories at "The USNS General Walker," 1st Battalion 22nd Infantry, http://1–22infantry.org/history3/walker.htm, and "General Nelson M. Walker," U.S. Navy Naval History and Heritage Command online, last modified July 10, 2015, https://www.history.navy.mil/research/histories/ship-histories/danfs/g/general-nelson-m-walker.html (accessed July 8, 2020).

9

Pavlics described his journey to Detroit in our interview of June 11, 2019. His assistance from the charity draws from Clow, "Law of the Stronger."

Peter Pavlics described his dad and Bekker in an April 15, 2020, email.

The Jupiter C–Juno I launch vehicle for *Explorer 1* is described in "Explorer-I and Jupiter-C: The First United States Satellite and Space Launch Vehicle" data sheet, NASA History online, https://history.nasa.gov/sputnik /expinfo.html (accessed August 20, 2020). For a detailed discussion of the respective roles of the U.S. Army and the U.S. Navy, see Neufeld, *Von Braun*.

Bekker's quote, and that in the following paragraph, is from his July 5, 1973, letter to Huntsville space historian Mitchell R. Sharpe (UAH).

Frank Pavlics described soil bins in our interview of June 11, 2019. See also Paramsothy Jayakumar et al., "Recommendation of a Next Generation Terramechanics Experimentation Capability for Ground Vehicle Systems," *Proceedings of the 7th Americas Regional Conference of the ISTVS* (2013).

10

For a discussion of rovers in science fiction, see Bettye B. Burkhalter and Mitchell R. Sharpe, "Lunar Roving Vehicle: Historical Origins, Development and Deployment," *Journal of the British Interplanetary Society* 48 (1995), 199–212.

The rest of the chapter relies on Hermann Oberth, *The Moon Car* (New York: Harper & Bros., 1959).

11

Bekker's quote is from his letter to Sharpe of July 5, 1973.

Pavlics described his work with scale models in our interview of June 11, 2019.

Bekker's move to GM was announced in "GM Eyes Mobility on Soils," *Detroit Free Press*, January 25, 1961, 17. The first half of his quote is from a letter he wrote to Mitchell R. Sharpe dated September 4, 1973; the latter half is from his letter to Sharpe of July 5, 1973 (UAH).

Pavlics described his move in our interview of June 11, 2019.

GM's Goleta campus still stands at 6767 Hollister Avenue. I visited the place, most of which is now occupied by a tech company called FLIR Systems, on June 12, 2019. FLIR's facilities manager, Tom Cattoi, shared several aerial photos of the campus from early in its GM days, and they're striking for the rural openness of Goleta and the skinny, doglegging Hollister Avenue.

GM's takeover of the place was reported in "GM Buys Plant in Goleta Valley," *Los Angeles Times*, August 19, 1960, 18; "Goleta Plant," *Santa Maria (Calif.) Times*, August 19, 1960, 2; and "GM Divisions Complete Move to Santa Barbara," *Los Angeles Times*, September 24, 1961, J-27.

My passage on Sam Romano's past relied on a short bio included in Boeing's LRV proposal of August 22, 1969, and telephone interviews with his family: with his wife, Marge, on April 5, 2020; his son Cliff on April 2, 2020; and his son Tim on April 3, 2020.

GM's loss of defense business and efforts to regain it were reported in Ray Moseley, "GM's Arsenal Takes Plunge," *Detroit Free Press,* August 16, 1959, 3. For more on the Defense Research Division's creation, see "New GM Division Created in Bid for Defense Jobs," *Detroit Free Press,* November 3, 1959, 3.

Pavlics told me in our interview of June 11, 2019, that he and Bekker almost immediately launched into rover studies. He's supported by press coverage: In "General Motors Researches Moon Exploration Vehicles," the *Van Nuys (Calif.) Valley News* reported that GM had tested three rover prototypes in its November 9, 1961, edition—well under a year after the company's arrival.

Jaquish's quote is from our telephone interview of July 27, 2020.

Calandro's quote is from our telephone interview of July 23, 2020.

Jaquish's comment about Romano's seriousness is from our telephone interview of July 27, 2020.

Cliff Romano made his remarks in our interview of April 2, 2020.

Marge Romano's quotes are from our interview of April 5, 2020.

Jaquish made his comments about Pavlics on July 27, 2020.

I interviewed Skaug on FaceTime on March 22, 2020, with his daughter, Betsy Doss, assisting.

Jaquish spoke about Bekker on July 27, 2020.

Marge Romano's quote about Bekker is from our interview of April 5, 2020.

Jaquish noted that Bekker did not take instruction on July 27, 2020.

Calandro spoke about Bekker in our telephone interview of March 5, 2020.

PART THREE: PRINCIPAL CONSIDERATIONS

12

Thomas Gold's paper "The Lunar Surface" appeared in *Monthly Notes of the Royal Astronomical Society* 115, no. 6 (1955). In it, he doesn't mention visiting spacecraft. The film I mention, *NASA Roundtable: Scientists Discuss the Moon,* can be viewed on YouTube, 21:28, Periscope Film, https://www.youtube.com/watch?v=xg4PFaZm4_k. An example of Gold doing himself no favors—or, alternatively, catching no breaks—can be found in "Dust on the Moon," *Time,* December 12, 1955. A short piece reporting on his aforementioned paper, it dutifully reports Gold's erosion theory until its last line: "Meantime, says Cosmologist Gold, spaceship pilots are advised not to land on the lunar plains."

Gold made the glacial comparison (and complained at length about how his position had been mischaracterized) in an April 1, 1978, interview for the American Institute of Physics, available on the AIP website at https://www.aip.org/history-programs/niels-bohr-library/oral-histories/4627 (accessed August 8, 2020). As for stories ascribing the quicksand idea to him, there's no shortage. NASA's own history websites make the claim (see https://www.hq.nasa.gov/office/pao/History/SP-4214/ch6-2.html (accessed August 13, 2020); Gold's own colleagues also make it (among them, Fred Whipple in

"The Coming Exploration in Space," *Saturday Evening Post*, August 16, 1958); and most of the obituaries that appeared after his June 2004 death figured it prominently—see, for example, Jeremy McGovern, "Thomas Gold, 1920–2004," *Astronomy* online, last modified June 24, 2004, https://astronomy.com /news-observing/news/2004/06/thomas%20gold%201920and1502004, and the tribute written by his friend and colleague Hermann Bondi, "Thomas Gold (1920–2004)," *Nature* 430, no. 6998 (July 22, 2004): 415, https://www .nature.com/articles/430415a.

Pavlics's description of early GM rover models is from our interview of June 11, 2019. Bekker's patent for the spaced-link track can be viewed at the Google Patents website, https://patents.google.com/patent/US2685481A/en (accessed August 8, 2020). His study of the "train concept" is outlined in his "Off-Road Locomotion on the Moon and on Earth," *Journal of Terramechanics* 3, no. 3 (1966), 83–91. Pavlics's quote about the model's agility, and the description of how its wheels helped one another, are from our interview of June 11, 2019. My passage on the partially flexible design was informed by Bekker, "Off-Road Locomotion on the Moon and on Earth." His later quotes are ibid.

The passage on GM's terrestrial vehicles was informed by "A Presentation on Roving Vehicles for Apollo Lunar Exploration Program," prepared by the Santa Barbara lab for NASA in May 1968 (UAH); *Vehicle Research & Development: GM Defense Research Laboratories,* a company booklet created in June 1965 to tout the Santa Barbara lab (UCSB); and Donald Friedman, "The Correlative Advantages of Lunar and Terrestrial Vehicle and Power Train Research," a paper presented at the Automotive Engineering Congress of January 10–14, 1966, and available for purchase through SAE Mobilus (pub. 660150).

13

The parade of early moon probe failures is cataloged on NASA's Solar System Exploration web pages; see https://solarsystem.nasa.gov/missions /?order=launch_date+desc&per_page=50&page=2&search=&fs=&fc=&ft=& dp=&category=. A terser tally of the heartbreak can be found in Ivan D. Ertel and Mary Louise Morse, *The Apollo Spacecraft: A Chronology,* vol. 1 (Washington, D.C.: NASA, 1969), https://www.lpi.usra.edu/lunar/documents/NTRS /collection3/NASA_SP_4009_1.pdf (both accessed September 11, 2020).

Von Braun's "a little less crash" remark is from "'Bugs' Persist in the U.S. Space Program," *Sydney (Australia) Morning Herald,* January 15, 1960, 2. The quote, worded differently, is also included in Andrew J. Dunar and Stephen P. Waring, *Power to Explore: A History of Marshall Space Flight Center, 1960–1990* (Washington, D.C.: NASA, 1999).

The Juno V's genesis is laid out in *Juno V Space Vehicle Development Program (Phase I)* (Huntsville: ABMA, October 1958), available online at https:// cdm16608.contentdm.oclc.org/digital/collection/p16608coll1/id/14281 /rec/1. The document notes that von Braun has proposed the name Saturn, though it has not yet been approved, and calls the rocket "the first real space vehicle, as the Douglas DC-3 was the first real airliner and durable workhorse in aeronautics."

Scholarship on Project Horizon is scant. About the only serious study of the subject is Frederick I. Ordway III, Mitchell R. Sharpe, and Ronald C. Wakeford, "Project Horizon: An Early Study of a Lunar Outpost," *Acta Astronautica* 17, no. 10 (1988), 1105–1121. The four-volume army report was later streamlined to two volumes for NASA, and while the original is hard to find, the latter is available for inspection online. Volume 1 at https://ns archive2.gwu.edu/NSAEBB/NSAEBB479/docs/EBB-Moon01_sm.pdf, lays out the army's argument for establishing the base and the political gain to be reaped by doing so. The lengthier volume 2 at https://nsarchive2.gwu .edu/NSAEBB/NSAEBB479/docs/EBB-Moon01A_sm.pdf, details how the army aimed to pull it off.

While both the *Acta Astronautica* paper and the report's second volume describe the vehicles I mention, they provide no further clues on the genesis of the ideas. One published account quotes GM's Sam Romano as saying the army approached the company for help, and that he recruited Greg Bekker and Frank Pavlics to GM for that purpose. That's almost certainly untrue: the project was secret and kept within the service; Romano wasn't even at GM at the time; and Bekker and Pavlics didn't join GM until long after. That said, the Horizon runabout so evokes Greg Bekker's thinking (as does the report's assertion that wheels would work just fine on the lunar surface, well before that became a popular view) that I'm moved to consider another possibility: that ABMA and Transportation Corps engineers working on Horizon conferred with Bekker's lab at Detroit Arsenal, which was within the same army command. At this writing, that remains a speculation; amid the disruptions of the Covid-19 pandemic, I have yet to find documentation that spells out such an arrangement.

Another possibility is that Bekker wasn't the only guy thinking as he did. Georg von Tiesenhausen, an ABMA and Marshall engineer whose papers are archived at the U.S. Space and Rocket Center, is widely credited with producing the first design of a lunar roving vehicle—in fact, he's often mistakenly said to have had a role in the eventual lunar rover, which he did not. But his papers include drawings, dated 1959, that roughly match the Horizon report's written description of the runabout.

My description of ABMA's transfer to NASA distills a lengthy and complex tussle into a single sentence. For more, see Michael J. Neufeld, "The End of the Army Space Program: Interservice Rivalry and the Transfer of the von Braun Group to NASA, 1958–1959," *Journal of Military History* 69, no. 3 (July 2005), 737–757; and Medaris, *Countdown for Decision*.

The decision to keep the von Braun team at Redstone is reported in "Von Braun's Team Will Stay on Site," *New York Times*, October 23, 1959, 12. See also Dunar and Waring, *Power to Explore*.

A facsimile of Von Braun's letter to LBJ is included in Roger D. Launius, *Apollo: A Retrospective Analysis* (Washington, D.C.: NASA, 1994). The book is available on the NASA Technical Reports Server at https://ntrs.nasa.gov/archive /nasa/casi.ntrs.nasa.gov/19940030132.pdf. See also Neufeld, *Von Braun*.

14

The Von Braun quote that opens the chapter is from his French Lick speech.

I interviewed Bill Vaughan by telephone on July 20, 2019.

My description of Marshall's organization relied on those sources, plus Neufeld, *Von Braun*, and Yasushi Sato, "Local Engineering and Systems Engineering: Cultural Conflict at NASA's Marshall Space Flight Center, 1960–1966," *Technology and Culture* 46, no. 3 (July 2005), 561–583. Von Braun's skills at board meetings was described by the same sources.

My description of von Braun's weekly notes was ibid.; Stephen B. Johnson, "Samuel Phillips and the Taming of Apollo," *Technology and Culture* 42, no. 4 (October 2001), 685–709; Neufeld, *Von Braun;* and Sato, "Local Engineering and Systems Engineering."

15

Marshall's shift from the "arsenal system" to contractor-based manufacture is described by Neufeld, *Von Braun,* and Sato, "Local Engineering and Systems Engineering."

I interviewed Al Haraway by telephone on January 3, 2020.

Von Braun's quote is from his French Lick speech.

16

Sonny Morea's life and career is drawn from our Huntsville interviews of April 24 and 25, 2019. We took our drive out to the "backyard" on April 25. I also made a longer, more detailed visit to the test stands with NASA's Craig Sumner on August 4, 2019.

For more on the Huntsville of 1955, see Carter, "Huntsville: Alabama Cotton Town Takes Off into the Space Age."

17

My passages on Prospector and Surveyor draw from "Lunar Program Underway," *Bulletin of the Atomic Scientists* 17, no. 2 (February 1962): 37–38.

The *Time* story I quote is "Free Enterprise v. the Moon," October 20, 1961.

The industry rovers are depicted in Wesley S. Griswold, "Weird Robots to Explore the Moon," *Popular Science*, March 1962, 96–99.

Bekker's reference to "science-fiction-like vehicles" is from his "Off-Road Locomotion on the Moon and on Earth." Subsequent quotes are from his

"Land Locomotion on the Surface of Planets." See also Bekker, "Mechanics of Locomotion and Lunar Surface Vehicle Concepts," a paper he presented at the Automotive Engineering Congress in January 1963, and available for purchase through SAE Mobilus.

The Bendix rover is described in Barry Miller, "Bendix Continues Own Lunar Rover Study," *Aviation Week & Space Technology*, June 4, 1962. I also relied on a November 7, 2019, phone interview with former Bendix staffer Sam Fine.

Bendix described its heavy investment in lunar mobility in *Lunar Surface Mobility Systems Technology Summary*, a report presented to NASA headquarters on May 28, 1968, and *Lunar Roving Vehicle Program Summary*, a report presented to NASA and congressional space committee staffers on October 30, 1968 (both UAH).

The GM reorganization was reported in "GM Moves Work of Defense Unit," *Oakland Tribune*, January 28, 1962, 33.

Jaquish's quote is from our interview of March 8, 2020.

The May 1963 GM paper I mention is titled "Lunar Roving Vehicle Concept: A Case Study" (staff paper SP-205) (UAH).

Marshall Space Flight Center engineer Otha "Skeet" Vaughn called my attention to the nineteenth-century wheel in two interviews: on June 26, 2019, at the Space and Rocket Center, and by telephone on August 26, 2019. He ascribed the design to an Englishman named Thomas Kitchens—and quoted Greg Bekker as having named Kitchens among the GM lab's influences. I subsequently encountered published references crediting the design to another Englishman, one Thomas Ricketts. While Skeet's testimony tilts me toward accepting Kitchens as the patent's author, I've been unable to verify it to my satisfaction, and thus struck it from the manuscript.

18

I'll admit that it took me awhile to "get" the harmonic drive, and I hope I've walked you through it effectively. If I've fallen short, I suggest you look up the Wikipedia page on "Strain wave gearing," where you'll find a simple GIF that puts the device in motion—and once you see that, it might make more sense.

One detail I didn't go into, in the interest of not overwhelming you: the harmonic drive is a flexible little device, in that you can hold the circular spline motionless, so that the flexspline moves, or—as in the case with the rover— you can hold the flexspline motionless, and make the moving circular spline part of the wheel.

Musser is the subject of a website, C. Walton Musser, http://www.walt musser.org. His patent is preserved at the Google Patents website, https://patents.google.com/patent/US2906143A/en (both accessed August 21, 2020).

Jaquish's quote is from our phone interview of March 8, 2020.

Pavlics's quote is from our interview of June 12, 2019.

19

Michael J. Neufeld offers a detailed analysis of the "mode" decision in "Von Braun and the Lunar-Orbit Rendezvous Decision: Finding a Way to Go to the Moon," *Acta Astronautica* 63, nos. 1–4 (July/August 2008): 540–50.

Von Braun's remarks, which he typed up after the conference, are included in Launius, *Apollo: A Retrospective Analysis.*

20

My description of the GM Surveyor rover draws from my interview with Pavlics on June 11, 2019, and "A Presentation on Roving Vehicles for Apollo Lunar Exploration Program," a slide show prepared by the Santa Barbara lab for NASA in May 1968 (UAH).

Bendix described its candidate in a *Final Technical Report* to the Jet Propulsion Lab, April 1, 1964. It's available for download on the NASA Technical Reports Server at https://ntrs.nasa.gov/citations/19660014483 (accessed September 4, 2020). See also the company's *Lunar Surface Mobility Systems Technology Summary* and *Lunar Roving Vehicle Program Summary* reports to NASA of 1968.

Calandro discussed the wire in our interview of March 5, 2020.

The showdown between the Surveyor rovers is described by Jack McCauley in Gerald G. Schaber, *The U.S. Geological Survey, Branch of Astrogeology—A Chronology of Activities from Conception Through the End of Project Apollo (1960–1973)* (Reston, Va.: U.S. Geological Survey, 2005), https://pubs.usgs.gov /of/2005/1190/of2005–1190.pdf.

My passage on the Ranger program was informed by NASA's Solar System Exploration website at https://solarsystem.nasa.gov/missions/?order=launch _date+desc&per_page=50&page=2&search=&fs=&fc=&ft=&dp=&category = (accessed September 11, 2020).

21

The Bendix report I quote is *ALSS: Scientific Mission Support Study* (March 1965), available at https://repository.hou.usra.edu/handle/20.500.11753/1108. The Bendix machine is described in detail in Richard E. Wong and Louis Galan, "Lunar Mobile Laboratory—Design Characteristics," a paper presented to the Automotive Engineering Congress in Detroit, January 10–14, 1966 (pub. 660146), and available through SAE Mobilus.

My description of GM's Molab was informed by "Surface Transportation Systems for Lunar Operations," a paper Sam Romano presented at the National Aeronautic and Space Engineering and Manufacturing Meeting in Los Angeles, October 4–5, 1965, and available from SAE Mobilus.

The "cold capsule" description was carried in a wire story about an August

1965 Molab demonstration. I found it under the headline "A Taxicab for the Moon," *Helena (Mont.) Independent-Record,* August 4, 1965.

Calandro's quote is from our interview of March 5, 2020.

Pavlics described the wheels' construction in our interview of June 11, 2019.

For more on the Lunex II experiment, see David Sheridan, "Two Emerge from 'Moon Rover' After 18 Days," *Minneapolis Tribune,* March 11, 1966, 19. The telephone booth reference is from Dick Caldwell, "'Moon Men' End 18-Day Test Without Losing Marbles," *Minneapolis Star,* March 11, 1966, 13A. That measurement is repeated in papers elsewhere, suggesting it came from NASA or Honeywell—though by my reckoning, Lunex II's volume was closer to four and a half phone booths.

I found some of Vaccaro's journal pages displayed on Case Antiques, an auction website advertising the July 2020 sale of his NASA memorabilia at https://caseantiques.com/item/lot-633-large-nasa-related-archive-inc-photographs-log-book/?utm_source=barnebys&utm_medium=referral&utm_campaign=barnebys&utm_content=2020–07–11 (accessed August 22, 2020). Vaccaro, by the way, would work on the LRV years later.

Vaccaro's "We were good friends" quote is from the documentary series *Moon Machines,* part 6, "The Lunar Rover," YouTube, 43:59, Chickstick, https://www.youtube.com/watch?v=J1TVj4TLYWo. Be advised that the doc isn't always reliable on the details.

Speaking of which, there are too many examples to cite of sources that misidentify the Mobile Geological Lab. The most recent I've seen is a story about GM mechanic Vic Hickey in the August 2020 issue of *Road & Track.*

22

My passage on Luna 9 was informed by an entry in the NASA Space Science Data Coordinated Archive, https://nssdc.gsfc.nasa.gov/nmc/spacecraft/display.action?id=1966–006A, and an article on RussianSpaceWeb.com that's particularly thorough, http://www.russianspaceweb.com/luna9.html. The description of Surveyor 1 drew from NASA's Solar System Exploration website (see https://solarsystem.nasa.gov/missions/? order=launch_date+desc&per_page=50&page=2&search=&fs=&fc=&ft=&dp=&category=), and NASA's Space Science Data Coordinated Archive, https://nssdc.gsfc.nasa.gov/nmc/spacecraft/display.action?id=1966–045A (all accessed September 11, 2020).

Romano described GM's proposed MTA in "Surface Transportation Systems for Lunar Operations." GM further described its machine in a 1971 pamphlet on the company's Santa Barbara activities (UCSB) and in "A Presentation on Roving Vehicles for Apollo Lunar Exploration Program."

Bendix described its MTA in the *Lunar Surface Mobility Systems Technology Summary* and *Lunar Roving Vehicle Program Summary* reports to NASA of 1968. See also Wong and Galan, "Lunar Mobile Laboratory—Design Characteristics."

The MTA tests at Aberdeen are described in "APG Testing 2 Concepts of NASA Lunar Vehicle," *Army Research and Development Newsmagazine,* October 1966, 8.

Von Braun's comments on a moon jeep are from "Dr. Wernher von Braun Tells How We'll Travel on the Moon," *Popular Science,* February 1964, 18.

The Brown Engineering test mule (for confusion's sake, also called a mobility test article) vanished for forty-plus years after the Apollo program, only to turn up rusting in a Blountsville, Alabama, backyard in the early 2010s. The careworn vehicle was eventually sold to Johnny Worley, a scrap dealer from Arab, Alabama, who twice offered it at auction as a "lunar rover prototype." It failed to find a buyer before his death in 2017 and is still for sale; interested parties should contact his son, Johnny Langston Worley Jr., of Hernando, Mississippi. (Telephone interview with the younger Worley, August 22, 2020.)

The LSSM and lunar shelter were not part of the Apollo Logistics Support System but a separate line of inquiry called the Apollo Extension System. Sam Romano recognized that although the LSSM "was derived by scaling down and simplifying the mobile laboratory design," it was likely to reach the moon first—and that, thus, the Molab "could be classified appropriately as a growth version of the local survey vehicle." For more, see Romano, "Surface Transportation Systems for Lunar Operations."

GM's LSSM was described further in its "A Presentation on Roving Vehicles for Apollo Lunar Exploration Program." The Bendix LSSM is described in the two presentation documents listed above. See also "LSSM Chronology," a timeline of the program prepared by Bendix (UAH).

The *Free Press* story I mention is in the newspaper's Sunday magazine: James H. Dygert, "Our Man in the Moon Car: What You Can Do in the 1970 Bendix," *Detroit,* May 7, 1967, 16. See also Larry Bush, "Newsman Earns His Driver's License for the Moon," *News-Palladium* (Benton Harbor, Mich.), April 17, 1967, 12.

23

My passages on the Lunar Wheel and Drive Experimental Test program relied on Vivake Asnani et al., "The Development of Wheels for the Lunar Roving Vehicle," NASA Glenn Research Center, December 2009 (TM 2009–215798); GM's "A Presentation on Roving Vehicles for Apollo Lunar Exploration Program"; my Calandro interview of March 5, 2020; and Norman J. James, *Of Firebirds & Moonmen: A Designer's Story from the Golden Age* (Bloomington, Ind.: Xlibris, 2007). Calandro described the bump stop's creation and James's contribution to the tread. See also J. K. Mitchell et al., *Materials Studies Related to Lunar Surface Exploration, Final Report,* vol. 3, a study prepared by the University of California Berkeley's Space Sciences Laboratory under NASA contract in March 1969, 1-A-10–13.

The tread's development is further explained in James, *Of Firebirds & Moonmen*.

Pavlics's quotes are from our interview of June 11, 2019.

24

My discussion of Ed Markow's career and contributions relied on telephone interviews with his family: his son, Jim, on April 1, 2020; his wife, Margaret, on April 1, 2020; and his daughter, Elizabeth Markow-Brown, on April 2, 2020.

For background on Markow's various wheel designs, see W. R. Herling and E. G. Markow, "Elliptical Wheel Concepts," a paper delivered at the International Automotive Engineering Congress, January 13–17, 1969, and available through SAE Mobilus. The metalastic wheel is explored further in E. G. Markow, "Predicted Behavior of Lunar Vehicles with Metalastic Wheels," a paper presented at the Automotive Engineering Congress of January 1963, and likewise available from SAE. Markow's patent is reported in a brief in the *Austin (Tex.) American*, May 16, 1965, 16.

The wheel is pictured on "Grumman Molab Concept," Unusual Off-Road Locomotion, https://www.unusuallocomotion.com/album/wheeled-robots /grumman-mtv.html (accessed September 3, 2020). Google "Grumman Molab," and you'll find umpteen other images. Incidentally, the craft inspired the moon mobile in a truly awful science-fiction movie, 1967's *Mission Stardust*, which you can watch on YouTube, if you dare. For more on the vehicle, see "Test-Run Is Given Model of Mooncar," *Record* (East Bergen, N.J.), November 30, 1965, 23.

I have not seen the Cronkite report I mention, but it's previewed with a photo of the newsman with Markow in "Life on the Moon," TV Week, *Valley Times* (North Hollywood, Calif.), February 3, 1967, 5. Markow's family also described it.

The Grumman LSSM is detailed in William Rice, "Grumman Shows a '72 Moon Rover," *New York Daily News*, October 4, 1968, 28. The conoidal wheel's performance on the jeep is described in Robert Gannon, "Driving a Go-Cart on the Moon," *Popular Science*, January 1971, 70. For a particularly satisfying description of craft and wheel, see Henry S. F. Cooper Jr., "Moon Car," *New Yorker*, January 18, 1969, 25.

The conoidal wheel is discussed in technical detail in *Design and Fabrication of Wheels for a Lunar Surface Vehicle*, a June 1970 Grumman report to NASA, available for download on the NASA Technical Reports Server at https://ntrs.nasa.gov/citations/19700022678 (accessed September 3, 2020).

The NASA concept I mention at chapter's end is illustrated in Anthony Young, *Lunar and Planetary Rovers* (Chichester, U.K.: Springer-Praxis, 2007), plate 61. Young indicates that it's a 1993 concept.

25

Philco's LunaGEM was described in "New Lunar Craft Planned," *Orlando (Fla.) Sentinel,* April 20, 1970, 4. See also "A GEM of a Moon Vehicle," *TODAY* (Cocoa Beach, Fla.), April 14, 1970, C10. The NASA response was included in the weekly notes of September 5, 1969.

The lunar worm was detailed in *Feasibility Study for Lunar Worm Planetary Roving Vehicle Concept,* a May 1966 report prepared under NASA contract by F. A. Dobson and D. G. Fulton of Philco. It's available on the NASA Technical Reports Server at https://ntrs.nasa.gov/search.jsp?R=19660022562 2020–07–29T02:20:31+00:00Z (accessed September 3, 2020).

The lunar leaper was described in Howard S. Seifert, "The Lunar Pogo Stick," *Journal of Spacecraft and Rockets* 4, no. 7 (July 1967): 941–43; William C. Harrison, "Scientist Proposes Giant Pogo Stick with People Pods," *Pittsburgh Post-Gazette,* February 3, 1967, 27; and George Getze, "Stanford Designing Pogo Stick for 50-Foot Hops Across Moon," *Los Angeles Times,* October 20, 1968, C1. More technical background can be found in Seifert and Marshall H. Kaplan, "Investigation of Hopping Transporter for Lunar Exploration," a June 1968 paper prepared under NASA contract, and Seifert et al., *The Lunar Hopping Transporter,* a July 1971 final report to NASA focusing on a much smaller leaper, both available for download on the NASA Technical Reports Server, https://ntrs.nasa.gov/search?q=hopping%20 transporter&highlight=true (accessed September 3, 2020).

As for the LFU, well, there's a ton of stuff out there. A good place to start is a blog called *No Shortage of Dreams* (formerly *DSFP's Spaceflight History*), which devotes a lengthy entry to flying belts and lunar flyers. The work of science writer David S. F. Portree, it's well worth your time; you'll find it at http:// spaceflighthistory.blogspot.com/2015/07/rocket-belts-and-rocket-chairs -lunar.html (accessed September 3, 2020).

A U.S. Geological Survey film about an August 1966 Bell rocket belt demonstration is available at https://www.facebook.com/watch/?v =2431004860289020. Punch "rocket belt" into YouTube, and you'll find a wealth of other clips.

The draft LFU program I mention is "Procurement Plan for the Development of a Lunar Flying Unit," prepared by MSC's Future Missions Project Office and dated (by its cover letter) October 13, 1967; the October 1968 comments on flyer concepts included one from MSC's Flight Control Division, dated October 11; and the draft statement of work for the LFU simulator was dated November 26, 1968, and was distributed by John D. Hodge, manager of the Advanced Missions Program, on December 5 (all NARA-FW).

For a detailed account of the North American Rockwell sit-down flyer, see *Study of One-Man Lunar Flying Vehicle,* its August 1969 final report to NASA. It and several other LFU documents are available for download on the

NASA Technical Reports Server at https://ntrs.nasa.gov/search?q=lunar%20 flying%20vehicle&highlight=true (accessed August 27, 2020).

PART FOUR: "WE MUST DO THIS!"

26

G. K. Grove's studies at Meteor Crater are described in Schaber, *U.S. Geological Survey.*

My passage on Daniel Barringer relied on Brandon Barringer, "Daniel Moreau Barringer (1860–1929) and His Crater," *Meteoritics* 2, no. 3 (December 1964).

Gene Shoemaker's story is told by Schaber, *U.S. Geological Survey.*

My passage on the creation of the Cinder Lake crater fields is ibid., and David A. Kring, "Cinder Lake Crater Field, Arizona: Lunar Analogue Test Site," a PowerPoint report at https://www.lpi.usra.edu/science/kring/lunar _exploration/CinderLakesCraterField.pdf. See also Amina Khan, "Astronauts and Geologists," *Los Angeles Times,* July 20, 2019, A2; Erika Ayn Finch, "Space Cowboys," *Sedona Monthly,* March 2018; and Lisa Messeri, "Earth as Analog: The Disciplinary Debate and Astronaut Training That Took Geology to the Moon," *Astropolitics* 12, nos. 2–3 (2014), 196–209.

Photos of the fields' creation and use are preserved in Geoff Manaugh and Nicola Twilley, "When We Blew Up Arizona to Simulate the Moon," *Atlantic* online, last modified January 23, 2013, https://www.theatlantic .com/technology/archive/2013/01/when-we-blew-up-arizona-to-simulate -the-moon/267456/. USGS video of the fields is available at the USGS Astrogeology Science Center online, https://astrogeology.usgs.gov/rpif/videos /making-craters (accessed September 11, 2020).

I visited them on October 10, 2019, and Meteor Crater on October 11, 2019.

27

The Summer Study report is available from the Hathi Trust, https://babel .hathitrust.org/cgi/pt?id=mdp.39015017174551&view=1up&seq=10.

28

For a thorough understanding of the lunar module, I'd direct you again to Woods's *How Apollo Flew to the Moon.*

That Bendix was actively studying the problem of fitting a rover into the lander is spelled out in the company's "Lunar Surface Mobility Systems Technology Summary" that it presented to NASA headquarters on May 28, 1968. "For the Extended Apollo Single Launch, the design of a vehicle smaller than either the LSSM or SP-LSSM is required," it wrote. Bendix foresaw a single-seat rover that would land "attached to the descent stage" and would have to be light.

Romano's family related the story of his trips to Grumman and NASA.

Pavlics described designing the rover's folding frame in our interview of June 11, 2019.

Pavlic's comments while showing me the model are ibid. His son Peter confirmed in an email of January 27, 2020: "My big contribution to the LRV was having my dad steal my Astronaut GI Joe to sit in the LRV model he built."

Pavlics described his meeting with von Braun in the June 11 interview.

29

Morea's quote at the chapter's opening is from a telephone interview on February 20, 2020.

Ben Milwitzky's letter of June 6, 1968, was to MSC's Andre J. Meyer (NARA-FW).

The October trip to Cinder Lake was mentioned in Schaber, *U.S. Geological Survey*.

The statement of work on dual-mode rovers was mentioned in a memo from James E. Saultz Sr., acting chief of the MSC's Experiments Systems Branch, to the chief of the Flight Control Division, dated March 18, 1969 (NARA-FW).

The Bendix quote is from its *Lunar Roving Vehicle Program Summary*.

The Grumman dual-mode rover is described in Robert De Piante, "Ride on Moon Is Simulated," *Cincinnati Enquirer*, February 8, 1970, 3-H.

GM's dual-mode rover designs are detailed in "Roving Vehicles for Apollo Lunar Exploration Program," a presentation dated May 1968 (UAH).

30

The April Management Council Meeting is mentioned in an "LRV Project Major Event Chronology," prepared by Morea in September 1969 (UAH).

My passage on the ballpark figure relied on the chart prepared by Marshall staff for NASA headquarters on May 5, 1969, delineating the cost estimates from Bendix, Grumman, and the center. Bendix figured it would cost $30 million in development, plus $3 million per rover; Grumman foresaw a total cost of $10 million to $20 million; and Marshall favored the figures that NASA went with.

See also "Chronology of NASA Cost Estimates Pertaining to Development and Production of Lunar Roving Vehicles," Summer 1971 (henceforth, "Cost Chronology"), an entry of which reads: "Based on [Marshall] estimates of approximately $30M for 3 vehicles, a figure of $40M for 4 vehicles was locked on and subsequently utilized in all conversations pertaining to LRVs between headquarters and [Marshall] during mid-1969."

Both of the above documents were included in an August 3, 1971, LRV cost growth analysis by Marshall's W. R. Adams (NARA-Atlanta).

The go-ahead telegram, signed by "William E. Stoney for Sam C. Phillips, Lt. General, USAF, Apollo Program Director," is preserved at UAH.

The follow-up memo from headquarters is referenced in "Cost Chronology."

Sonny Morea described his Montreal trip and phone conversations in our interviews of April 24–25 and June 26, 2019. Months later, as I was going through papers I'd scanned at UAH, I came across handwritten notes he'd made in the wake of the job offer. In the main, they listed the pros and cons of accepting, and there were more of the latter. Morea had doubts about the assignment's effects on his career (he saw it as a lateral move, at best) and on the Engine Program Office (which he noted was already suffering from poor morale). What grabbed my attention, however, was a scrawled notation at the top of the first page: "LRV Job offer came 3:20 P.M. PDT while I was in L.A.—3 June 69. The discussion came the next morning at 8:30 A.M., 4 June 69." I wrote Morea for some clarification on whether he was perhaps misremembering the Montreal part of the story, attaching the scan of his notes, and we discussed that possibility by phone on June 16, 2020. He told me he was certain that he was in Montreal, and on examining the document concluded that it must refer to a separate, later conversation in which the details of the job were hammered out. I am noting this here, lest future researchers encounter the document and wonder what's up.

31

The procurement regimen is laid out in several drafts (with the dates slipping later in each new round) on file both at UAH and NARA-Atlanta. The statement of work is likewise filed at both.

Transportation routes taken by the Saturn V's stages are depicted in Ashley Morrow, "An Unexpected Journey: Spacecraft Transit the Panama Canal," NASA Marshall Flight Center online, last modified April 9, 2015, https://www.nasa.gov/content/goddard/an-unexpected-journey-spacecraft-transit-the-panama-canal/. Morea discussed the impetus for the Michoud pitch in a telephone interview on February 20, 2020.

The twenty-nine companies invited to bid were identified in a letter to those firms from contracts officer Elbert "E. B." Craig on June 27, 1969 (UAH).

Drafts of the request for proposals are filed at UAH. The solicitation made clear that the cost of preparing a bid was on the bidder; the winning company could not apply that cost to the resulting contract.

Marshall hosted a bidders' conference at Michoud on July 23, 1969. "The purpose of this conference," Morea explained in the weekly notes, "was to answer any questions on the RFP and to give prospective bidders an opportunity to look over the Michoud facilities as a possible site for the performance of this contract." Attending were reps from North American Rockwell, Boeing,

General Motors (AC), TRW, Westinghouse, Bendix, Allis-Chalmers Manufacturing Company, Grumman, and Chrysler (NARA-Atlanta).

One aspect of the conference is particularly interesting. An entry in the "Cost Chronology" notes that Marshall officials were "concerned that the $40M figure being utilized within NASA might unduly influence cost proposals in process of being submitted by various companies bidding for the LRV effort.

"As a result," the entry reads, "companies attending the bidders' conference . . . were informed that NASA intended to hold costs down wherever possible on the LRV project and that bids should reflect their companies' honest judgment without being biased by any cost estimate figure they may have previously heard discussed within NASA."

32

Boeing's Al Haraway described the HIC Building in our interview of January 3, 2020.

The quote from Boeing is from *Lunar Roving Vehicle Technical Proposal D5– 17013,* Boeing Huntsville, August 22, 1969 (henceforth, Boeing proposal).

Pavlics's quote on GM's tepid support is from our June 11, 2019, interview. Romano made his comment in the *Moon Machines* documentary.

Pavlics told me he figured they were a "shoo-in" in our June 11, 2019, interview.

The quotes re von Braun at Grumman are from Cooper, "Moon Car," 25.

33

The construction of the Apollo EMU is described in detail by ILC Industries of Frederica, Delaware, the subcontractor for much of the suits, in "Space Suit Evolution: From Custom-Tailored to Off-the-Rack," Human Space Flight section on NASA online, last modified 1994, https://spaceflight.nasa.gov /outreach/SignificantIncidentsEVA/assets/space_suit_evolution.pdf. The article also includes lots of interesting insight into suits used for Skylab and Shuttle missions.

Craig Sumner's quotes are from a FaceTime interview on July 23, 2019.

Charlie Duke described an astronaut's limited visibility in a telephone interview on August 27, 2020.

Jerry Schaber, who mapped the Apollo 11 traverses, described the limited range of the crew's travels in an interview at his home in Flagstaff on October 11, 2019.

The white-knuckle moments before Apollo 11's landing are so well known it seems silly to cite a source here, but if you haven't seen the amazing 2019 CNN Films picture *Apollo 11,* here's your kick in the pants.

The July 29, 1969, debriefing was described in the weekly notes of August 4, 1969. The Apollo 11 crew made the twenty-foot-wheel suggestion in an

August 7, 1969, meeting chronicled by George M. Low, then the manager of MSC's Apollo Spacecraft Program, in an August 13 memo (NARA-FW).

Morea reported additional soil studies, and planning for a demonstration at Cinder Lake, to Marshall Center management in a slide report on August 15, 1969 (UAH).

The September 1969 trip to Cinder Lake is described in Morea's input to the weekly notes, dated October 6, 1969.

Incidentally, the astronauts were not alone in expressing doubt about the LRV. A September 9, 1969, briefing by Morea (UAH) indicates that the Manned Spacecraft Center's director, Robert R. Gilruth, challenged the need for the craft and "the entire lunar exploration concept." This manifested a long-standing rivalry between the Marshall Center and Houston—mostly between their top-level administrators rather than working engineers. As Michael Neufeld thumbnailed it for me, Marshall was responsible for the three stages of the Saturn V, and Houston for everything above, including all parts of the craft that involved the astronauts; Houston thus saw Marshall as invading its territory by developing what amounted to a manned spacecraft for use on the moon's surface.

34

The bulk of this chapter is drawn from the Boeing proposal.

See also Henry Kudish and Sam Romano, "Dual Power Steering Maneuvers Lightweight Platform Chassis," *SAE Journal* 78, no. 5 (May 1970). The article, which described the Boeing/GM proposal in detail, was part of a large package of stories on the LRV; another, by NASA's Ben Milwitzky, laid out the agency's requirements and explained what it hoped to gain with a rover. I obtained the journal through SAE Mobilus.

35

My description of the Bendix proposal was informed by "Preliminary Mockup Indicates Design Approach Used by Bendix," *SAE Journal* 78, no. 5 (May 1970). The story was prepared by the journal's editors "with the cooperation of Bendix engineers" but did not draw from the proposal itself, as it "was not released." The newspaper story I mention is Hugh McCann, "Weird Car Begins Race to Moon in a Quarry," *Detroit Free Press,* August 16, 1969, 1.

My passage on the Chrysler proposal relied on C. C. Gage and W. S. Parker, "Back-to-Back Astronauts Ride on Chassis Much Like a Car's" in the same *SAE Journal* edition. Gage was Chrysler's manager on the program, and Parker its engineering manager.

My description of the Grumman proposal was informed by another story in that May 1970 issue of *SAE Journal:* "Motorized Body Articulation Steers Soft Conical Wheels," written by Ed Markow himself. The *New Yorker* quote is from Cooper, "Moon Car," 25.

I gleaned further details of the Bendix, Chrysler, and Grumman proposals from Sonny Morea's evaluations of them, contained in charts and handwritten notes dated "September 1969" and "Sensitive" (UAH).

36

The Source Evaluation Board was described after its first meeting in a July 11 letter from its chairman, James E. Bradford. In it, he listed the criteria the board would use. They were further enumerated in an undated memo marked "Sensitive" (both NARA-Atlanta).

Thomas Paine's remarks spelling out the SEB's findings were contained in a letter to the Marshall Center from NASA headquarters, incorporating two statements by Paine regarding LRV contract selection and dated November 16, 1970 (UAH). His "Statement 1" covered this material.

For Morea's assessment, see notes for chapter 35. It's worth noting that under the heading "Incentives," he wrote: "Obviously, Boeing has gotten the message of our philosophy of we win they win we lose they lose—no one else comes close."

Paine described meeting with the board and NASA executives on September 29, 1969, and his decision, in his November 16, 1970, Statement 1.

Morea's quote is from our interview of April 24, 2019.

Jim Markow's quote is from our interview of April 1, 2020.

SEB's questions and the answers from Boeing are filed at NARA-Atlanta.

Marshall's Propulsion and Vehicle Engineering Lab signaled its preference for Bendix's nutator speed-reduction system in its monthly progress report for March 1968, after GM had completed its lunar wheel study and Bendix had finished a separate inquiry into the nutating drive system: "The results of the drive system experimental test programs completed by AC Electronics and Bendix were evaluated," it read. "As a result, it was recommended that preference be given to the nutator drive mechanism and that the harmonic drive be developed as backup."

The incentive structure is spelled out in Exhibit C of the Boeing contract; in an undated slide presentation Morea made on the contract's structure (UAH); and in an October 14, 1969 summary, also prepared by Morea (NARA-Atlanta).

Negotiator Kenneth M. "Mike" Grant's comments about Bendix are from "Summary of Negotiations—Letter Contract," October 22, 1969 (UAH).

Paine's account of the contract award to Boeing is from his "Statement 2," included in the headquarters letter of November 16, 1970. Morea's quotes on the subject are from our interview of April 24, 2019. The letter contract is filed at NARA-Atlanta.

Morea told me several times that he was concerned from the start that Boeing's bid undershot the project's actual expenses. The "Cost Chronology" indicates how concerned he was: it says that his office tried to keep $40 million as the project's cost in its budgetary docs but was told by headquarters it couldn't do that. Instead, in December 1969 it submitted a program operating plan estimate, or POP, of $32.2 million for the LRVs, which "represented a genuine concern for a probable cost escalation . . . in the neighborhood of 50 to 60 percent of the prime contractor estimates."

Mike Grant told me about the Bendix rep's reaction in a telephone interview on September 9, 2020.

Frank Pavlics recalled the celebration at GM in our interview on June 11, 2019. Marge Romano remembered the occasion, too: "When it was announced, we had gone out to celebrate at a little local restaurant in Goleta at the time," she told me on April 5, 2020. "They were all on top of the moon."

PART FIVE: A PAINFULLY TRYING TASK

37

This chapter is entirely the product of hindsight, but a detailed breakdown of Boeing's crowded to-do list can be gleaned from the statement of work dated December 30, 1969 (NARA-Atlanta).

38

Morea raised the lunar communications relay unit weight issue in his input to the weekly notes dated October 10, 1969. Regarding use of the LCRU in case of a rover breakdown, see my notes for chapter 54.

Morea's letter to Milwitzky about weight was dated October 24, 1969 (UAH). He mentioned the letter and his concerns about weight in his input to the weekly notes dated October 24 and November 7, 1969.

Hauessermann's concerns were reported in Morea's answers to questions raised in the weekly notes of December 18, 1969.

That Marshall favored a brushless, permanent magnet motor is suggested by Clyde S. Jones Jr. et al., "Traction Drive System Design Considerations for a Lunar Roving Vehicle," a 1969 paper by NASA engineers that I obtained through SAE Mobilus. That thinking hasn't changed within the agency. When I interviewed the Johnson Space Center's William Bluethmann on the upcoming VIPER lunar rover, he told me that GM's reliance on a brush motor "kind of blows my mind."

My description of Marshall's culture of accountability relied on Johnson, "Samuel Phillips and the Taming of Apollo"; Sato, "Local Engineering and Systems Engineering"; and my interview with Bill Vaughan on July 20, 2019.

Morea describes his thoughts on the nav system, and the Avionics Lab's concurrence, in his submission for the same mid-December weekly notes. My description of the nav system was informed by "The Navigation System of the Lunar Roving Vehicle," a technical memorandum prepared by Bellcomm—a think tank formed from Bell Labs to provide technical advice to NASA—on

December 11, 1970, and available through the ALSJ at https://www.hq.nasa
.gov/alsj/19790072520_1979072520.pdf.

The TV camera business was spelled out in an Engineering Branch note to
Morea on January 6, 1970 (UAH).

The backgrounds of Kudish and Newman are included in the Boeing proposal. I also relied on my Morea interview of February 20, 2020, and my
Haraway interview of January 3, 2020.

The first project review is described by Boeing in its monthly progress
report, no. 2.

Boeing's price hikes were chronicled in "Memorandum of Negotiation,
Definitive Contract NAS8–25145," a document prepared by contract officer
Mike Grant on January 26, 1970. His "most stringent discipline" comment
is ibid. (UAH). See also "Addendum No. 1 to Pre-Negotiation Conference
Report, Dated January 12, 1970," also by Grant (UAH).

39

Apollo 20's cancellation is explained in John Uri, "50 Years Ago: NASA Can-
cels Apollo 20 Mission," NASA History online, last modified January 3, 2020,
https://www.nasa.gov/feature/50-years-ago-nasa-cancels-apollo-20-mission.
George Low made his comment in the course of denying rumors that the last
four Apollo missions were on the chopping block. He was quoted by UPI; see
"Lunar Landings to Be Cut," *Philadelphia Inquirer,* January 5, 1970, 1.

Morea's note to his staff was dated January 12, 1970 (UAH). Roy Godfrey
issued his warning in the weekly notes of January 19, 1970.

For more on Von Braun's departure for headquarters, see Neufeld, *Von
Braun;* Harold M. Schmeck Jr., "Von Braun to Go to Washington to Direct
Space Mission Plans," *New York Times,* January 28, 1970, 26; and Dunar and
Waring, *Power to Explore.* Morea spoke of von Braun's influence in an inter-
view on June 26, 2019.

My description of Rees drew from Dunar and Waring, *Power to Explore.*

My passage on the PDR drew from an LRV program status report dated
January 14, 1970 (UAH), and Boeing's monthly progress report no. 3, issued
February 20, 1970. Neither the Boeing report nor any other documentation
I've encountered mentions the elimination of the roll bar. In an email of Sep-
tember 5, 2020, retired Marshall engineer Ron Creel told me it was "before
or at the PDR."

My description of the wax tanks relied on Creel's emails of August 31,
September 2, and September 5, 2020. Creel worked on the LRV's thermal
subsystem.

The Vomit Comet flights were described by Jerry Carr in a Skype interview
on December 12, 2019. I also relied on Morea's input for the weekly notes of
January 23, 1970; the fifty-nine parabolas are noted in the weekly notes of
March 30, 1970.

40

Morea explained his worries about missed milestones in our interview of February 20, 2020. His comment to Kudish, and the response, are ibid.

Boeing suggested that the program's slippage was at least partly NASA's doing in its second monthly progress report, dated January 10, 1970. NASA was annoyed by this and told the company so in a phone call on January 22, 1970. That prompted a letter dated January 26, 1970, from J. M. Neal, Boeing's LRV program control manager. "It was not the intent of Boeing to indicate that NASA . . . had generated all of the changes," he wrote, "but only to point out that as a result of the changes, some time was lost" (NARA-Atlanta).

In our phone interview of February 20, 2020, Morea told me that Kudish had assured him that Boeing would catch up. His worries about Boeing's communication with GM is ibid., and from an undated "Chronology" of the project's schedule and budget woes (hereafter, "Morea Chronology," which Morea evidently prepared as a memo for the record) (UAH). His comments about GM's research character is from our interview of February 20, 2020.

Morea quoted Kudish as saying, "Yeah, you're right," in the same interview.

Morea recorded his February 12, 1970, comments to Romano in a handwritten memo of the same date (UAH). He summarized the meeting in the "Morea Chronology." The reference to the 1G trainer is from Morea's "Memorandum for Record: Rationale for Government's Decision to Approve the Initial Boeing Company's Purchase Order with General Motors on April 2, 1970," evidently written months later (and henceforth "Rationale") (UAH).

Morea's "not the slightest glimmer" quote is from handwritten notes he made after a program review on February 26, 1970. He called the review a "fiasco" in notes he prepared the next day, after another meeting with Kudish (UAH). Morea followed up those exchanges with Kudish in a letter dated March 2, 1970, in which he reiterated his concerns about the lack of program control (UAH).

He expressed empathy for Kudish in our interview of February 20, 2020.

The "1–3 month slip" warning (which Morea also aired in his March 2 letter to Kudish) was part of his February 27 input to the weekly notes.

Morea noted his creation of a tiger team in the "Morea Chronology."

41

My passage on contractor penetration was informed by Sato, "Local Engineering and System Engineering," and Dunar and Waring, *Power to Explore.* Morea's quote is from our interview of February 20, 2020.

The "hip pocket" judgment is from a summary attached to a memo dated April 14, 1970, from W. R. Adams to the LRV project's James Belew (NARA-Atlanta). The tiger team's other findings were included in handwritten notes Morea kept during the team's review on March 16, 1970 (UAH). The "weak link" remark was included in the weekly notes of April 6, 1970.

Petrone's call is mentioned in the "Morea Chronology." Godfrey reported that Boeing was dispatching a team in the April 6, 1970, weekly notes. Morea noted that AC had sent a team to Santa Barbara in his April 10 input to the weekly notes; in handwritten notes dated April 17, he wrote further: "AC Milwaukee has recognized how the mobility system can go 'down [the] tube' unless they send some help. Plan is to layer key experienced people on present organization from Milwaukee" (UAH).

Grant's memo of March 18, 1970, concerning the Boeing/GM contract is filed as a weekly activity report at UAH.

42

The rover's weight is listed in Boeing's monthly progress report no. 5, released April 20, 1970. The vibration unit's status is ibid. The company's push to delay the CDR is ibid.

Morea's "I finally got frustrated" quote is from our interview of February 20, 2020. According to the "Morea Chronology" entry for April 15–17, 1970, Morea "pointed out need to [Boeing] and GM for overlaying some experienced project and program control people on present organization." The next entry, for April 17, noted "complete revamping of organization."

Haraway's "Earl was an old iron banger" quote is from our interview of January 3, 2020.

Cowart's quote is from an interview at his home on August 3, 2019.

43

The frame mishap at GM—which occurred on April 29, 1970—was detailed in a letter from two of the GM engineers involved to Sam Romano, dated April 30 (NARA-Atlanta). Morea's description of the incident is from his input to the weekly notes, dated May 1, 1970.

Boeing's letter to GM, dated May 12, 1970, is filed at NARA-Atlanta.

Boeing's "areas of concern" are referenced in E. B. Craig's response, dated May 21, 1970 (NARA-Atlanta).

Morea apparently got advance word of Boeing's overrun announcement, as signaled by his input to the weekly notes, dated May 22, 1970: "I am informed that the contractor is reassessing his total cost picture and plans to deliver an unscheduled report on May 25 projecting a significant project overrun." The formal announcement came, as promised, three days later. Morea offered a thorough analysis of Boeing's announcement in a June 1970 special status report for NASA managers (NARA-Atlanta).

Lee James reported on the impossibility of Boeing's numbers in a letter to Eberhard Rees on June 19, 1970 (NARA-Atlanta). He invoked the C-5A debacle in the same letter.

The do-or-die meeting of June 18, 1970, was detailed in Morea's memo for the record, dated June 19. The document is the source for all the comments I've quoted (NARA-Atlanta).

Boeing's price boost is from the "Cost Chronology."

Rees's letter to Stoner was dated June 23, 1970; Stoner's reply was dated July 1, 1970 (both NARA-Atlanta).

The CDR on the mobility subsystem—the part GM was building—was conducted May 26, 1970, "with a minimum of new action items," according to the weekly notes of June 1, 1970. Boeing reported results of the June 16–17 CDR on the rest of the rover in its monthly progress report no. 7, issued on June 24.

Craig's letter to Boeing was dated June 24, 1970 (NARA-Atlanta).

44

Greg Bekker's retirement, effective June 1, 1970, was announced in "Movers & Shakers," *Detroit Free Press,* May 31, 1970, 25.

His "perhaps I have been a precursor" quote is from his letter to Mitchell Sharpe on July 5, 1973.

Don Beattie's communication with Putty Mills, and the creation of the Grover, were detailed by both men in Schaber, *U.S. Geological Survey.* I also relied on an interview with Mills facilitated by his daughter, Denice Dogan, in November 2020. Mills, who was ninety-six at the time, was sufficiently hard of hearing that he couldn't be interviewed by telephone, and couldn't be visited due to Covid-19. Instead, using questions that I'd posed in an email, Dogan chatted with him by computer, then passed along his answers.

Schaber described Mills and the Grover during our interview of October 11, 2019.

Astronaut training in the Grover was detailed in Schaber, *U.S. Geological Survey.*

Duke's comments regarding the division of duties is from our interview of May 3, 2019. His comments about learning to play travel guide are from our interview of October 30, 2019.

The Grover's roles sans astronauts are detailed in Schaber, *U.S. Geological Survey.*

45

Morea identified August 1970 as the project's most imperiled month in "Rationale."

The rover's weight was reported in the weekly notes of July 13, 1970. Boeing reported damage to the vibration unit, the halt in battery testing, the cancellation of the engineering test unit, and delays in the qualification test unit in its monthly progress report no. 9 (which Boeing mistakenly labeled no. 8).

Problems with the torsion bars were noted in Morea's input to the weekly notes, dated July 20, 1970, and in a letter from Boeing Huntsville to E. A. Schmidt of AC/DRL on August 14, 1970 (NARA-Atlanta).

Issues with the steering motors and harmonic drives were reported in the weekly notes of June 29, 1970; in Morea's input to the weekly notes, dated July 24; and in the weekly notes of September 21. Lee James raised pulling the steering in-house in an undated memo to Rees about a July 2 meeting with Dale Myers (henceforth referred to as "Undated Lee James Memo to Rees" (NARA-Atlanta).

Bekker described the February 3, 1970, meeting with Armstrong, Aldrin, and Apollo 12 astronaut Alan Bean in his letter to Sharpe of July 5, 1973. He appeared to take the moonwalkers' comments in good humor. Not so those from the Corps of Engineers: "I vividly recall GM and Boeing engineers resenting, [sic] the unwarranted and perhaps unfair criticism by the engineers of the Waterways Experiment Station," he wrote to Sharpe on August 24, 1973. "This antagonism was, in the opinion of a number of people, the source of unnecessary cost and tension."

The Corps's modeling at WES was described by Marjorie Anders in an Associated Press report that I found as "Corps of Engineers Lab Is Disaster Think Tank," in the *Carlsbad (N. Mex.) Current-Argus* of May 26, 1983, 8.

The wheel's difficulties were listed by Morea in his submission for the weekly notes dated October 24, 1969. Experiments with fabric coverings are ibid. That climbing a 25-degree slope was essential is noted not only in the statement of work but also in Sam Romano's work diaries, which cover the period from October 31, 1969, to April 26, 1971, in two volumes, and were lent to me by his son Tim. Romano's entries for December 2, 1969 (vol. 1, 14) include a telephone conversation with Henry Kudish. His takeaway: "The wheel must climb 25-degree slope, or it is not suitable for LRV. Other req may not be as important."

Complaints about the original joystick apparently started months before the August 10–12, 1970, design review at which the issue was flagged as critical. Houston classified a change as mandatory on September 11, 1970; the edit was noted in the weekly notes of September 21. By October 6, when Saturn Program boss Richard G. Smith made a status report on the LRV project to headquarters, the new design was in hand (UAH).

The astronauts' test of the new design took place on October 27, 1970, per the November 2 weekly notes.

46

Dale Myers evidently suggested a cost ceiling in a telephone call to Rees July 27, 1970. It is referenced in Rees's response, dated July 28, 1970, by teletypewriter (NARA-Atlanta).

Morea's "break the $30 million mark" quote is from his July 31, 1970, input to the weekly notes.

Rees's "never ending" letter was dated August 25, 1970 (NARA-Atlanta).

In building Romano's background, I drew on interviews with his family and his bio in the Boeing proposal.

Paul Blasingame's letter to Rees was dated September 21, 1970 (NARA-Atlanta). The assessment that GM had misjudged the task was evidently shared by its partner. In an memo for the record, dated August 19, 1970, NASA's C. F. Reynolds wrote that he met with Boeing officials, who "seemed to feel that [GM], being unfamiliar with NASA standards, rating requirements, and qualification standards just didn't realize the problems involved. [GM] apparently had no reason to train their people to comply with NASA standards since not [sic] usually necessary in their work" (NARA-Atlanta).

Pavlics commented on "unreal" paperwork in our interview of June 12, 2019.

The Marshall assessment, ordered by Saturn Program chief Richard G. Smith, was reported in Morea's "Rationale."

Marge Romano's quote is from our interview of April 5, 2020.

The Romano diary entries for August and September appear in vol. 2, which covers the period from July 23, 1970, to the program's end.

The AC-Delco merger is chronicled in "Delco Is Merged with AC Electronics," *Kokomo (Ind.) Tribune*, September 1, 1970, 1; "Delco Electronics Division Shifts Executives," *Kokomo (Ind.) Tribune*, September 3, 1970, 2; and "GM Recognition of Kokomo," editorial, *Kokomo (Ind.) Tribune*, September 3, 1970, 4.

Rees's letter to Myers was dated August 31 (NARA-Atlanta).

Boeing's proposal to move its LRV operation to Kent is chronicled in the slide presentation used at the August 25 meeting and a report distributed to attendees, and also in an August 28 memorandum for the record about the meeting by Saturn deputy program manager Brian O. Montgomery (NARA-Atlanta).

Morea spelled out his reservations in an August 28 letter to Houtz. The reply, dated September 10, earned margin notes from Morea: next to the paragraph pushing back on cost savings, he scrawled to Montgomery: "Monty!! Comment?"

47

A nice summary of the Apollo program's shrinkage can be found in Thomas O'Toole, "Flying to the Moon: How Many Times?" *Washington Post*, December 10, 1972, C1.

Duke's comment on Apollo 15 getting the first rover is from our interview of May 3, 2019.

The small savings from canceling LRV-4 was mentioned in the "Undated Lee James Memo to Rees." LRV-4's use as a parts car is detailed in a letter dated December 2, 1970, from Joe Jones of Marshall's Test and Operations Branch, and in requests for contractual action dated January 25 and February 17, 1971 (NARA-Atlanta).

Grant offered his program summary in a memorandum for the record dated October 5, 1970 (NARA-Atlanta).

Boeing's price boost to $33.4 million is recorded in the entry for October 21,

1970, in the "Cost Chronology." "In addition," the entry reads, "changes valued at \$526K were identified bringing the cost to \$33.9M."

Boileau made his appeal in an October 26 bulletin to the Boeing Aerospace Group staff (NARA-Atlanta).

48

This chapter relied most heavily on A. P. Vinogradov et al., eds., *Lunokhod-1: Mobile Laboratory on the Moon* (Foreign Technology Division translation, U.S. Air Force Systems Command, November 1971), a Soviet publication translated by the Foreign Technology Division at Wright-Patterson Air Force Base, Ohio. It is the most complete account of Lunokhod-1's design and capabilities that I found (UAH). See also Mikhail Malenkov, "Self-Propelled Automatic Chassis of Lunokhod-1: History of Creation in Episodes," *Frontiers of Mechanical Engineering* 11, no. 1 (2016): 60–86; Tomas de J. Mateo Sanguino, "50 Years of Rovers for Planetary Exploration: A Retrospective Review for Future Directions," *Robotics and Autonomous Systems* 94 (2017) 172–185; and Giuseppe De Chiara and Michael H. Gorn, *Spacecraft: 100 Iconic Rockets, Shuttles, and Satellites That Put Us in Space* (Minneapolis: Voyageur Press, 2018).

The teletypewritten message of December 31, 1970, regarding Low's trip to the Soviet Union was signed by Erich Neubert, the Marshall Center's acting deputy director, but was sent by Morea (NARA-Atlanta).

49

Morea enumerated the project's stumbles in a memo to Rees, prior to the latter's December 4 call to Stoner (NARA-Atlanta). Problems with the 1G trainer and continuing slips in the qualification test unit were mentioned in Morea's November 6, 1970, input to the weekly notes. Issues with the batteries were flagged in the weekly notes of November 16.

Rees's letter to Stoner, notes from Stoner's December 6 huddle with his LRV team, and Stoner's December 8 letter to Rees are filed at NARA-Atlanta.

Pavlics recalled the day of celebration at Santa Barbara and told me of his first drive in our interview of June 12, 2019.

Morea announced the 1G trainer's arrival at MSC—with the new hand controller installed—in his input to the weekly notes dated December 18. Its use by the astronauts was reported in Morea's input for January 21, 1971.

Boeing's price increase to \$36.5 million is recorded in the January 13, 1971, entry to "Cost Chronology." The later jump to \$37.8 million is ibid., in the entry for February 5.

The "spiraled" quote is from a letter dated February 12, 1971, written by contracts office head O. M. Hirsch (NARA-Atlanta).

Morea noted in his February 5, 1971, input to the weekly notes that new price negotiations had started.

Morea told me of the effects of the *Post* story, and his call from Milwitzky, in our interview of April 25, 2019. His "busy squealing" quote is from our interview of February 20, 2020; his closing quote about not hearing from Petrone is from our April 25 interview.

50

Morea mentioned issues with the wax tank attached to the drive control electronics in his November 19, 1970, input to the weekly notes. He noted electronic interference in his inputs dated February 12, 19, and 26, 1971. The torsion bar issue is mentioned in Engineering Branch notes to Morea dated February 22 (NARA-Atlanta).

The completion of testing at Pismo was noted in the July 6, 1970, weekly notes. GM experimented with tread coverage from March 19 to April 6, 1970, and recommended the 50 percent design; it was approved by NASA on May 1, according to an LRV program review conducted at Santa Barbara on May 6 (NARA-Atlanta). The original design's call for aluminum tread is in the Boeing proposal.

Morea offered his judgment that the wheel was ready in his March 12, 1971, input to the weekly notes.

The wheel chatter was first mentioned in the weekly notes of September 21, 1970. Morea mentioned meetings to address the issue in his September 25, October 15, October 22, and October 29 inputs to the notes. In his entry dated November 13, Morea noted: "Due to the LM payload weight problem, MSC is striving to minimize this subject as a problem." In his input dated March 19, 1972, he reported that the chatter had disappeared.

Boeing's original space support equipment (SSE) was described in its proposal. The company's struggles getting it to work were signaled in its monthly progress report no. 7, in which it said it would redesign it. Morea reported July 10, 1970, for the weekly notes that it again required redesign, which Boeing seconded in its monthly progress report no. 8, wherein it noted "deficiencies encountered during deployment testing." Cowart's comments are from our interview of August 3, 2019.

The weekly notes of August 3, 1970, reported a new "semiautomated" design and nineteen additional pounds. Boeing reported its success in testing in its monthly progress report no. 10. Morea's "do not give us the confidence" quote is from his September 25 input for the weekly notes. His October 2 input noted continuing "concern about the reliability of the SSE." In addition, astronaut Robert Parker filed a lengthy September 25 report about deployment failures during a days-long run-through at Grumman (NARA-FW).

Rees's "sleepless nights" comment was in an October 26 memo to Saturn Program boss Richard G. Smith (NARA-Atlanta). The weekly notes of December 14 detailed continuing difficulties, including the first saddle break. The notes of December 21 reported the second saddle break. The notes of January 11, 1971, mentioned still more problems but floated the idea of a manual replacement. Those of January 18 said that while repairs were under way, "we have directed Boeing to proceed on a parallel backup effort to develop and qualify a manual deployment system." Morea's January 21 input for the notes reported success with the manual gear. He announced the cancellation of the semiautomatic version in his February 12 input.

Morea was finally able to report "No further problems are foreseen with SSE" in his input of April 1, 1971.

My description of the deployment, as well as of the instructions, were informed by the *LRV Operations Handbook*, available on the ALSJ website at https://www.hq.nasa.gov/alsj/lrvhand.html.

51

Duke described practicing in the Grover and the 1G trainer in our interview of October 30, 2019.

My description of the Pogo is ibid.

Craig Sumner described the LRV simulator in our interview of July 23, 2019; Otha "Skeet" Vaughn also detailed it in our interview of August 26, 2019.

Sumner described the Vomit Comet flights in our interview of July 23, 2019. Duke related the need to lie down during the plane's dives in our interview of October 30, 2019.

52

Morea reported completion of the qualification tests in his February 26, 1971, weekly notes input. The seat belt changes are ibid.

Morea reported LRV-1's delivery to the government in his March 12 input to the weekly notes. Its delivery to Kennedy Space Center was reported in Boeing's monthly progress report no. 16 and in Morea's weekly notes input of March 19, 1971.

Finished rover weights were provided me by retired Marshall engineer Ron Creel in an August 31, 2020, email. LRV-1 tipped the scale at 464.71 pounds, according to Creel's figures; with the SSE, it weighed 494.25; and loaded in the lunar module, it and its associated gear came to 508.75 pounds. With LRV-2, Boeing got the rover down to 462.12 pounds, but the other figures budged less than a pound. With LRV-3, the weights ballooned: the rover alone weighed 471.49 pounds, and loaded aboard the LM, it totaled a whopping 518.9 pounds.

Morea learned of the fender droop in a February 22 Engineering Branch report (NARA-Atlanta); the new fenders' balkiness, and the labs' solution, were noted in Morea's April 9 input to the weekly notes. Scott's flag request was noted in a Flight Control Division report dated March 31, 1971, about a recent crew station fitting session (NARA-FW).

The battery regimen was described in Morea's March 5, 1971, input for the weekly notes.

PART SIX: ACROSS THE AIRLESS WILDS

53

Shepard's golf shot can be viewed online: "Alan Shepard Playing Golf on the Moon During Apollo 14," YouTube, 1:36, Science News, https://www.youtube.com/watch?v=qdrcRGxDxHQ.

The $8 million figure was cited by the Associated Press and other wire services, and turned up in scores of papers; see, for example, "Moon Buggy Most Expensive Car Ever Built," *Courier-Express* (DuBois, Pa.), July 31, 1971. A few outliers cited other figures; for instance, the *Detroit Free Press* priced each rover at $3.5 million, and the *New York Times,* noting that most of the project's expense was in research and development, figured each at $2 million.

Not only did the *Times* devote nearly a full inside page to the rover, but much of its July 31 front page—and the machine was likewise big news in papers from the *Los Angeles Times* (banner, end-of-the-world-size, all-caps headline: "First Moon Ride"), to the afternoon *St. Louis Post-Dispatch* ("Scott and Irwin Drive Buggy on the Moon"), to the *Fairbanks (Alas.) Daily News-Miner* ("Scott and Irwin Drag Across Man in the Moon").

The Schlosser's ad appeared in the *Daily Leader (Pontiac, Ill.),* July 24, 1971, 3.

Pavlics described the launch in our June 12, 2019, interview; Marge Romano described it in our April 5, 2020, interview.

My description of the landing site was informed by the ALSJ; the *Apollo 15 Mission Report* (Houston: Manned Spacecraft Center, December 1971), available at https://www.hq.nasa.gov/alsj/a15/ap15mr.pdf; Apollo 15 EVA maps, from NARA-Atlanta and NARA-FW; and contemporary news coverage of the mission, especially that of the *New York Times.* Most helpful of all were the lunar reconnaissance orbiter camera photos of the site, shared online by Arizona State University. The entire LROC selection of images is worth exploring, but most helpful here were EVA traverse maps at http://lroc.sese.asu.edu/posts/491 and oblique views of the area at http://lroc.sese.asu.edu/posts/362. NASA offers up a helpful analysis using the photos at https://www.nasa.gov/content/goddard/apollo-15-original-interplanetary-mountaineers (all accessed August 28, 2020). See also "Geologic Setting of the Apollo 15 Samples," *Science* 175 (January 28, 1972), 407–15.

Dave Scott's quote during the stand-up EVA is from the ALSJ at https://www.hq.nasa.gov/alsj/a15/a15.seva.html.

Pavlics's reminiscence is from our interview of June 12, 2019.

The steering loss is documented on the ALSJ at https://www.hq.nasa.gov/alsj/a15/a15.lrvdep.html.

Romano's reaction is from his *Moon Machines* appearance.

Pavlics's reaction is from our June 12 interview.

The Flight Control Division's mission rules, dated April 5, 1971, are filed at NARA-FW.

The ALSJ records Irwin mounting the rover and his seat belt trouble at https://www.hq.nasa.gov/alsj/a15/a15.trv1prep.html. The start of their first drive is at https://www.hq.nasa.gov/alsj/a15/a15.elbowtrv.html.

54

Background on Dave Scott draws from his memoir, *Two Sides of the Moon* (New York: Thomas Dunne Books, 2004), which he cowrote with Soviet cosmonaut Alexei Leonov; the ALSJ; and his NASA bio at https://www.nasa.gov/sites/default/files/atoms/files/scott_david.pdf.

For Jim Irwin's background, I relied on his NASA bio at https://www.nasa.gov/sites/default/files/atoms/files/irwin_james.pdf; an AP story that detailed his boyhood and air crash, and was published as "The Three Men Going to the Moon: James B. Irwin Jr.," *Des Moines (Iowa) Register,* July 26, 1971, 4; and his obituary, John Noble Wilford, "James B. Irwin, 61, Ex-Astronaut; Founded Religious Organization," *New York Times,* August 10, 1991, 26.

Apollo documents suggest that the contingency plans in the event of a rover breakdown almost certainly included hand carrying the lunar communications relay unit, without which the astronauts would be out of contact with Houston until they came within range of the lunar module's radio relay. The LCRU could be set up to draw power from its own batteries or from the rover—a design detail that allowed for its use independent of the vehicle. The LCRU training manual details how it is to be prepared for a walking traverse. (See the doc on the ALSJ website at https://www.hq.nasa.gov/alsj/HSI-481184-LCRU.pdf.) Likewise, NASA's lunar surface procedures manual for Apollo 15 provides instructions for its use when removed from the LRV (see https://www.hq.nasa.gov/alsj/a15/a15lsp.pdf); it shows that in addition to being hand carried, the fifty-four-pound box could be strapped to the lunar module pilot's backpack.

Those documents notwithstanding, Dave Scott told me unequivocally that the LCRU would not have been used on a walk back. "'We' (the crew and the flight crew support team) developed and verified the final flight procedures used for each system and each operational situation," he wrote in an email on June 27, 2020. "During the preparation for Apollo 15, carrying the LCRU was never even considered."

Charlie Duke told me that he, too, had no memory of such a plan: "I don't recall even discussing this option, much less training for that contingency," he wrote in a July 7, 2020, email.

So we're left with a contradiction: the LCRU's documentation on one side, and two guys who actually used the device on the other. But seeing as how the lunar module's role as a relay station was limited by range and its reliance on line-of-sight communication, and given the distances and terrain that might have separated a stranded crew from base, I'm willing to bet that with or without training, Houston would have insisted that the crew set out for home lugging the LCRU. I ran this past ALSJ creator Eric Jones, who sees it the same way.

Boeing claimed a range of fifty-seven miles in *Lunar Roving Vehicle,* a booklet it created for use by the press prior to the Apollo 15 launch (UAH). NASA claimed just a forty-mile range in its Apollo 15 press kit, available on the ALSJ at https://www.hq.nasa.gov/alsj/a15/A15_PressKit.pdf.

Charlie Duke explained the farthest-first strategy in our interview of May 3, 2019.

Scott's "eye on the road" comment appears on the ALSJ at https://www.hq.nasa.gov/alsj/a15/a15.elbowtrv.html. His first "Whoa!" is ibid. So is his second. Irwin's "bronco" comparison is ibid. The exchange about the rille is ibid.

Scott's initialization of the nav system is included on the ALSJ at https://www.hq.nasa.gov/alsj/a15/a15.trv1prep.html.

Geology at Elbow Crater is chronicled on the ALSJ at https://www.hq.nasa.gov/alsj/a15/a15.elbow.html, and the drive to Hadley Delta at https://www.hq.nasa.gov/alsj/a15/a15.trvsta2.html.

My passage on the subsequent stop draws from the ALSJ at https://www.hq.nasa.gov/alsj/a15/a15.sta2.html.

That account of the drive back to the lander relies on the ALSJ at https://www.hq.nasa.gov/alsj/a15/a15.trvlm1.html.

55

The opening scene and drive to Station 6 draws from the ALSJ at https://www.hq.nasa.gov/alsj/a15/a15.trvsta6.html.

Their first stop is chronicled on the ALSJ at https://www.hq.nasa.gov/alsj/a15/a15.sta6abv.html and https://www.hq.nasa.gov/alsj/a15/a15.sta6crtr.html. My estimates of the crew's elevation are based on NASA's analysis of LROC photos, and available at https://www.nasa.gov/content/goddard/apollo-15-original-interplanetary-mountaineers.

The passage at the green boulder relied on the ALSJ at https://www.hq.nasa.gov/alsj/a15/a15.sta6a.html.

The Spur Crater scene draws from the ALSJ at https://www.hq.nasa.gov/alsj/a15/a15.spur.html.

Griffin's quote was published in several slightly different forms. I used the version published by, among others, *TODAY,* the hometown paper of Florida's space coast—now known as *Florida Today*—and altered the punctuation so that the quote made better sense. See Sanders LaMont's "The Greatest Scientific Mission of All Time," the lead story in the August 8, 1971, edition.

Comments during the trip back to the lander were chronicled on the ALSJ at https://www.hq.nasa.gov/alsj/a15/a15.trvlm2.html.

56

Scott's frustrated quotes draw from the ALSJ at https://www.hq.nasa.gov/alsj /a15/a15.coreextract.html.

The drive is chronicled on the ALSJ at https://www.hq.nasa.gov/alsj/a15 /a15.trvsta9.html; the stop at the crater at https://www.hq.nasa.gov/alsj/a15 .sta9.html; and the arrival at the rille at https://www.hq.nasa.gov/alsj /a15/a15.rille.html.

The ALSJ breaks down the final, brief stop at https://www.hq.nasa.gov /alsj/a15/a15.sta10.html, and the drive back to the lander at https://www .hq.nasa.gov/alsj/a15/a15.trvlm3.html.

My description of the theater at EVA's end relied on the ALSJ at https:// www.hq.nasa.gov/alsj/a15/a15.clsout3.html. Some readers will note the terrible irony of Scott's postal demonstration, for the Apollo 15 astronauts were destined to get into a storm of trouble over stamps. For more, see Harold M. Schmeck Jr., "Apollo 15 Crew Is Reprimanded," *New York Times,* July 12, 1972, 1; and Howard Muson, "Comedown from the Moon: What Has Happened to the Astronauts," *New York Times Magazine,* December 3, 1972, 37.

Video of the hammer and feather drop is available at Chris Higgins, "Hammer and Feather Drop on the Moon," Mental Floss, https://www.mentalfloss .com/article/22913/hammer-and-feather-drop-moon, accessed August 30, 2020.

Dave Scott's activity while parking the rover is chronicled on the ALSJ at https://www.hq.nasa.gov/alsj/a15/a15.clsout3.html. No sooner had the stamp flap died down than Scott found himself at the center of another public dustup, when the Belgian artist responsible for the tiny aluminum "Fallen Astronaut" figurine started turning out copies intended for sale at $750 a pop—in violation, Scott said, of the terms through which he got the moon commission. NASA wasn't pleased. For details, see Howard Muson, "Fallen Astronaut," *New York Times Magazine,* December 3, 1972, 139; and Corey S. Powell and Laurie Gwen Shapiro, "The Sculpture on the Moon," on the Slate website at http:// www.slate.com/articles/health_and_science/science/2013/12/sculpture_on _the_moon_paul_van_hoeydonck_s_fallen_astronaut.html.

Footage of *Falcon*'s liftoff can be found at "Apollo 15 Lunar Liftoff (Inside and Outside View)," YouTube, 4:08, Apollo Houston, https://www.youtube .com/watch?v=JmBc2y_m-2A. An inset shows the view from inside the ascent stage; about a minute into the flight, there's a great view of the rille.

Joe Allen's report on the TV camera is documented on the Apollo Flight Journal, a companion to the ALSJ that offers transcripts of everything spoken aboard the Apollo spacecraft while they were under way. Created by W. David Woods, it affords all of us earthbound types a peerless view of life inside the capsules. This exchange is preserved at https://history.nasa.gov/afj /ap15fj/24day12_presser.html.

The Astrionics Lab's inquiry into the LRV's performance was detailed in the weekly notes of August 2, 1971. LRV-1's performance was also summarized in Nicholas C. Costes et al., *Mobility Performance of the Lunar Roving Vehicle: Terrestrial Studies—Apollo 15 Results* (Washington, D.C.: NASA, December 1972—Technical Report N73–16817/TR R-401). The "MEF" crack is from an August 26, 1971, postflight report by experiments officers Gerald Griffith and Richard Koos (NARA-FW).

Morea's comments are from his input to the weekly notes dated August 6, 1971.

Lee Scherer's comment was reported in "Success of Moon Rover May Broaden Future Apollo Plan," *Tampa Tribune*, August 2, 1971, 1.

Sam Romano's son Tim told me of the GM party in our interview of April 3, 2020. He later mailed me a copy of the invitation.

57

Dave Scott's "kiddie bar" quote is from the Apollo 15 technical debriefing, available on the ALSJ at https://www.hq.nasa.gov/alsj/a15/a15tecdbrf.html.

The testing of the replacement seat belts was detailed by Morea and his chief engineer, Jim Sisson, in several editions of the weekly notes in the fall of 1971.

My account of Apollo 16's travels to the moon is from the *Apollo 16 Mission Report*, available on the ALSJ at https://www.hq.nasa.gov/alsj/a16/A16_MissionReport.pdf.

The description of the landing site relied on ibid; from Arizona State's LROC website at http://lroc.sese.asu.edu/posts/529; and from my interview with Duke of May 3, 2019.

Pavlics's quote is from our interview of June 12, 2019.

Background on Young relied on the memoir he wrote with James R. Hansen, *Forever Young* (Gainesville: University Press of Florida, 2012); his NASA bio at https://www.nasa.gov/sites/default/files/atoms/files/young_john.pdf; and Andrew Chaikin, *A Man on the Moon* (New York: Viking Penguin, 1994).

Since I've mentioned the Apollo 15 reprimands in the notes for chapter 56, it's only right that I note that Young, too, was reprimanded—for sneaking a corned beef sandwich aboard Gemini 3. An hour and fifty-two minutes into that flight, he pulled it from his suit and offered half to his commander, Virgil "Gus" Grissom. "Where did that come from?" Grissom asked.

"I brought it with me," Young answered. "Let's see how it tastes." After a few bites, Grissom complained that his half was falling to pieces and shoved it into his pocket. "Pretty good, though," he said, "if it would just hold together."

After a long pause, Young asked: "Want some chicken leg?"

NASA had an institutional stroke.

I gleaned Duke's background from the memoir he wrote with his wife, Dotty—*Moonwalker* (Nashville: Thomas Nelson, 1990); from his NASA bio at https://www.nasa.gov/sites/default/files/atoms/files/duke_charles.pdf; and from our interview of May 3, 2019.

The crew's exits from the lander are transcribed on the ALSJ at https://

www.hq.nasa.gov/alsj/a16/a16.eva1prelim.html. Its deployment of the rover and Young's test ride are ibid at https://www.hq.nasa.gov/alsj/a16/a16.lrvdep.html.

Young's comment about driving with or without steering is from the Apollo 16 technical debriefing, available on the ALSJ at https://www.hq.nasa.gov/alsj/a16/a16tecdbrf.html.

My passage on the first drive to Flag and Plum Craters, and science there, relied on the ALSJ at https://www.hq.nasa.gov/alsj/a16/a16.trvsta1.html and https://www.hq.nasa.gov/alsj/a16/a16.sta1.html. Duke's comments to me are from our interview of May 3, 2019.

The second stop is detailed on the ALSJ at https://www.hq.nasa.gov/alsj/a16/a16.sta2.html.

My account of the grand prix relied on the ALSJ at https://www.hq.nasa.gov/alsj/a16/a16.trvlm1.html, and on video of the event at https://www.youtube.com/watch?v=X30z82aeSHw. When Eberhard Rees learned of the upcoming event in the weekly notes of January 17, 1972, he annotated them with: "What the 'H . . .' is the Grand Prix objective?" Saturn Program boss Richard G. Smith provided the answer in a January 28 memo: "This is terminology that has been applied to the LRV driving test on the lunar surface. . . . This provides accurate data on the LRV dynamics, wheel bounce, wheels off the ground, wheel slip, LRV side slip in turns, and dust trajectory. None of this can be obtained except in actual 1/6-g environment."

Duke's closing quote is preserved on the ALSJ at https://www.hq.nasa.gov/alsj/a16/a16.clsout1.html.

58

All passages on the second EVA were informed by the ALSJ: the first drive and science stop at https://www.hq.nasa.gov/alsj/a16/a16.trvsta4.html and https://www.hq.nasa.gov/alsj/a16/a16.sta4.html, respectively; the downhill run and Station 5 at https://www.hq.nasa.gov/alsj/a16/a16.sta5.html; the trouble en route to Stubby at https://www.hq.nasa.gov/alsj/a16/a16.trv6to8.html; Young's encounter with the fender at https://www.hq.nasa.gov/alsj/a16/a16.trv6to8.html; and the resulting dust shower at https://www.hq.nasa.gov/alsj/a16/a16.trvlm2.html.

Young's remarks are from the technical debriefing. Duke's comments are from our telephone interview of October 30, 2019.

The scene at Pavlics's house is taken from Bekker's letter to Mitchell R. Sharpe of July 5, 1973.

59

My account of the third EVA relied on the ALSJ. Transcripts of the traverse to North Ray Crater are at https://www.hq.nasa.gov/alsj/a16/a16.trvsta11.html; the science stop there, and at House Rock, at https://www.hq.nasa

.gov/alsj/a16/a16.sta11.html and https://www.hq.nasa.gov/alsj/a16/a16.house
_rock.html, respectively; the run down North Ray and stop at Shadow Rock
at https://www.hq.nasa.gov/alsj/a16/a16.trvsta13.html and https://www.hq
.nasa.gov/alsj/a16/a16.sta13.html; and the EVA's end and LRV-2's final parking
spot at https://www.hq.nasa.gov/alsj/a16/a16.trvlm3.html and https://www
.hq.nasa.gov/alsj/a16/a16.vip.html.

Duke's comment to me about House Rock is from our interview of May 3,
2019.

Duke's yell on takeoff is preserved on the ALSJ at https://www.hq.nasa
.gov/alsj/a16/a16.launch.html.

60

Young's and Duke's comments are from the Apollo 16 technical debriefing.

Sam Romano's efforts to drum up interest in adding remote control to
LRV-3 are chronicled in his LRV diary, vol. 2. It indicates that Romano discussed the idea with NASA officials first on November 17, 1970—the day of
Lunokhod-1's landing—and over the next several months met or spoke about
it several times with Ben Milwitzky and other agency executives, as well as
with people at Boeing. On April 21, 1971, he met with Sonny Morea; Morea
"was impressed," Romano wrote, but warned him not to spend any LRV project money on it. Romano met with officials of RCA on April 26. After that,
Romano wrote nothing, but evidently the idea survived at least through the
summer: in his September 9, 1971, submission for the weekly notes, Morea
wrote that Boeing, GM, and RCA reps met with MSC officials about it over
Labor Day.

Earlier in the year, the Manned Spacecraft Center in Houston mulled
equipping Apollo 17 with a small remote-controlled rover, in addition to
LRV-3. The craft, its four silicone-rubber wheels driven by "chains and sprock-
ets," would weigh 350 pounds. That idea, too, came to naught.

My account of the fender's destruction and Cernan's repair were informed
by the ALSJ at https://www.hq.nasa.gov/alsj/a17/a17.alsepoff.html.

My description of the landing site and the crew's planned travels relied
on the *Apollo 17 Mission Report*, at https://history.nasa.gov/alsj/a17/A17
_MissionReport.pdf; the ALSJ; EVA maps and timetables in the LRV col-
lection at UAH and the LRV files at NARA-Atlanta; an essay, "The Valley of
Taurus-Littrow," at https://www.hq.nasa.gov/alsj/a17/a17.site.html; and Eugene Cernan and Don Davis, *The Last Man on the Moon* (New York: St. Martin's Press, 1999). I also found a small amount of connective tissue regarding
the mission in capcom Robert Parker's papers at NARA-FW.

The crew's first rover outing is chronicled on the ALSJ at https://www
.hq.nasa.gov/alsj/a17/a17.trvsta1.html, at https://www.hq.nasa.gov/alsj/a17/a17
.sta1.html, and at https://www.hq.nasa.gov/alsj/a17/a17.trvlm1.html.

I assembled Schmitt's background from his NASA bio, available at https://www.nasa.gov/sites/default/files/atoms/files/schmitt_harrison.pdf.

My account of activities back at the lander relied on the ALSJ at https://www.hq.nasa.gov/alsj/a17/a17.clsout1.html.

61

My description of Apollo 17's launch was informed by Chaikin, *A Man on the Moon;* Cernan and Davis, *Last Man on the Moon;* the *Apollo 17 Mission Report;* and by footage of the event, available at "Launch of Apollo 17 (TV Feed and NSA Footage)," YouTube, 7:56, lunarmodule5, https://www.youtube.com/watch?v=rmwc8E9fCLI.

Cernan's background is drawn from his NASA bio at https://www.nasa.gov/sites/default/files/atoms/files/_cernan_eugene_a._deceased_pdf_75_kb_.pdf, and from Cernan and Davis, *Last Man on the Moon.*

Young's "simple-minded procedure" comments are reserved on the ALSJ at https://www.hq.nasa.gov/alsj/a17/a17.eva2wake.html. The repair is chronicled at https://www.hq.nasa.gov/alsj/a17/a17.outcam.html.

The dinner with von Braun is recounted in Cernan and Davis, *Last Man on the Moon.*

The epic journey to Station 2 is covered on the ALSJ at https://www.hq.nasa.gov/alsj/a17/a17.outcam.html and https://www.hq.nasa.gov/alsj/a17/a17.outcam.html.

The walk-back limits are laid out on the ALSJ at https://www.hq.nasa.gov/alsj/a17/a17.sta2.html.

A double failure was discussed in a July 1971 risk assessment by the NASA Aerospace Safety Advisory Panel, available at https://history.nasa.gov/asap/1971-Apollo15.pdf. "One particular case of double failure was noted and questioned by the panel, i.e., possibility of LRV and PLSS failure at the same time," it reads. "MSC indicated that the probability of such a double failure was extremely remote. Although structural failure of the LRV is considered a hazard, testing and analysis appear to have made this highly unlikely and consequently an acceptable risk."

The science at Station 2 is chronicled at https://www.hq.nasa.gov/alsj/a17/a17.sta2.html.

My account of the traverse to Station 3 and the science at Ballet Crater relied on the ALSJ at https://www.hq.nasa.gov/alsj/a17/a17.trvsta3.html and https://www.hq.nasa.gov/alsj/a17/a17.trvsta4.html, respectively.

The ALSJ captures the excitement of Schmitt's orange soil discovery at https://www.hq.nasa.gov/alsj/a17/a17.sta4.html. See also G. M. Brown, J. G. Holland, and A. Peckett, "Orange Soil from the Moon," *Nature* 242 (April 1973).

The drive to Camelot Crater, and the science there, was chronicled on the

ALSJ at https://www.hq.nasa.gov/alsj/a17/a17.trvsta5.html and https://www.hq.nasa.gov/alsj/a17/a17.sta5.html, respectively.

Cernan's thank-you to John Young is recorded on the ALSJ at https://www.hq.nasa.gov/alsj/a17/a17.clsout2.html.

The Auto Body Association of America honor was reported in "Body Men Honor Apollo 17," *Asbury Park (N.J.) Press*, December 14, 1972, 1.

My passage on the post-EVA conversation in the lander relied on the ALSJ at https://www.hq.nasa.gov/alsj/a17/a17.eva2post.html.

62

My description of the crew's preparations for the last day's drive, its journey to the first station, and the science it performed there, relied on the ALSJ at https://www.hq.nasa.gov/alsj/a17/a17.trvsta6.html and https://www.hq.nasa.gov/alsj/a17/a17.sta6.html.

Troctolite 76535 is described at https://curator.jsc.nasa.gov/lunar/lsc/76535.pdf.

My account of the Station 8 stop was informed by the ALSJ at https://www.hq.nasa.gov/alsj/a17/a17.sta8.html.

The drive to Station 9 and the action at Van Serg were chronicled on the ALSJ at https://www.hq.nasa.gov/alsj/a17/a17.trvsta9.html and https://www.hq.nasa.gov/alsj/a17/a17.sta9.html, respectively.

Kestay's quote is from our crater fields visit of October 10, 2019.

My description of the rover's parking and cleanup relied on the ALSJ at https://www.hq.nasa.gov/alsj/a17/a17.clsout3.html.

Schmitt's poem is preserved on the ALSJ, for better or worse, at https://www.hq.nasa.gov/alsj/a17/a17.launch.html.

The crew's liftoff can be viewed at "Apollo 17 Liftoff from Moon—December 14, 1972," YouTube, 0:36, Smithsonian National Air and Space Museum, https://www.youtube.com/watch?v=9HQfauGJaTs (all accessed August 27, 2020).

PART SEVEN: TIRE TRACKS

63

Von Braun's opening quote is from Al Rossiter Jr., "Space Pioneer Envisions U.S. Shelters and Gardens on Moon," *Arizona Republic* (Phoenix), December 3, 1972, B3. His second quote is from an essay he wrote himself, "Space Pioneer Reflects on Apollo's Achievements," *New York Times*, December 3, 1972, 68.

Donald Bickler's invention of the rocker-bogie was described in Bettuane Levine, "Out of this World," *Los Angeles Times*, December 15, 1996. I was also fortunate enough to have an email exchange with Bickler on August 25, 2020, in which he mentioned his work with Pavlics. For his part, Pavlics told me of his work as a consultant during Sojourner's development through his son, Peter, in a September 11, 2020, email.

My passage on Sojourner relied on Bickler, "Roving over Mars," *Mechanical Engineering*, April 1998, and "The Pathfinder Microrover," *Journal of Geophysical Research* 102, no. E2 (February 25, 1997): 3989–4001.

Bekker's quote is from his letter to Sharpe of July 5, 1973. Bekker's belated recognition in Poland was reported by Joseph Reaves in "Poland Now Honors Lunar Rover Designer," *Chicago Tribune*, October 22, 1991, 22.

64

Of all the stuff the astronauts left behind, the only things the public expressed much interest in having back were the rovers. In the days after the Apollo 15 crew's departure from Hadley Base, NASA received a cascade of offers for what the press called "the most expensive used car in the solar system." Bob Mears, the owner of the Space Age Alarm Company of Bakersfield, California, appears to have made the highest bid: $8,000, provided that NASA arranged for its delivery.

Bob Gilruth, the Manned Spacecraft Center's director, wrote to Mears that the rover had not yet been declared government surplus. "I am sure if the vehicle would be offered for sale," he added, "its selling price would have to take into consideration its low mileage, its excellent condition, and also the fact that it is some 250,000 miles from Earth." In other words, Mears would have to do a lot better than $8,000. "But," Gilruth closed, "you will certainly be contacted if it ever becomes available to the public." See "Manned Space Center Rejects Offer for Rover," *St. Petersburg (Fla.) Times*, August 24, 1971, 7A.

My discussion of the lunar reconnaissance orbiter is based on an email exchange with ASU's Mark Robinson on June 13, 2020. It was also informed by J. W. Keller et al., "The Lunar Reconnaissance Orbiter Mission: Six Years of Science and Exploration at the Moon," *Icarus* 273 (2016); R. V. Wagner et al., "Coordinates of Anthropogenic Features on the Moon," *Icarus* 283 (2017); and I. Haase et al., "Mapping the Apollo 17 Landing Site Area Based on Lunar Reconnaissance Orbiter Camera Images and Apollo Surface Photography," *Journal of Geophysical Research* 117 (2012).

The LRO spent its first two years photographing the moon from about thirty-one miles up. In 2011 its orbit was tightened to about sixteen miles, and it captured the sharp images to which I refer. "The average size of an LRO [camera] pixel is fifty centimeters—twenty inches—so there is no way we could identify actual boot prints," Robinson told me. "But it was clear from watching the movies and videos of the astronauts on the moon that they were quite messy when walking about, so it was possible we would see disturbances of the regolith . . . from orbit."

That any pictures were possible is an almost unbelievable achievement because, as Robinson explained it, the narrow-angle cameras are "made of carbon fiber, a hydrophilic material" that plumps as it absorbs water and then shrinks as it dries out. "You have to build the telescope out of focus, so as it desiccates and shrinks in the space environment, the structures that hold the lenses shrink, and the lenses move to the in-focus position. So I was a bit worried until we got our first set of images down." I encourage you to pause for a moment to let that sink in.

65

Morea guessed at the rovers' present condition in our interview of April 25, 2019.

Pavlics made his assessment in our interview of June 12, 2019.

Skeet Vaughn offered his judgment in our conversation of June 26, 2019.

Pavlics's "the kids didn't see me" quote is from our June 12 interview.

Gene Cowart made his comment during our interview of August 3, 2019.

Morea's "I marvel" quote is from our interview on February 20, 2020.

Morea's "lessons learned" memo to Rees, stamped "Sensitive" on every page, is filed at UAH. The date of the document is illegible, but a note by program management chief Lee James, attached to copies he was forwarding to other Marshall bosses, is dated December 24, 1970. Some of Morea's comments made James and others bristle, judging from their comments, while others resonated. One recipient, identity unknown, wrote: "In giving Morea the LRV proj., we gave him an impossible job" (UAH).

Morea's "Otherwise the astronauts" comment is from our interview on June 26, 2019.

Bekker's comments are from his letter to Sharpe dated August 24, 1973.

Cowart made his "You go around Huntsville" comment on August 3, 2019.

I saw Pavlics's signed photo during my visit of June 11, 2019.

66

My passage on the VIPER rover draws from a telephone interview with William Bluethmann on June 22, 2020. See also Kenneth Chang, "NASA Hires Pittsburgh Firm to Take Rover to the Moon," *New York Times*, June 12, 2020, B8.

The scene aboard *Challenger* that closes the chapter relied on the ALSJ at https://www.hq.nasa.gov/alsj/a17/a17.eva2post.html (accessed September 11, 2020).

INDEX